A RIVER NAMED LEE

A RIVER NAMED LEE

CALYAN
PUBLISHING LTD.

VIVIENNE POY

First published in Canada by

Calyan Publishing Ltd.
4151 Sheppard Avenue East, 2nd floor
Scarborough, Ontario
Canada M1S 1T4

ISBN 1-896501-00-1

Poy, Vivienne
A River Named Lee

English Editor: Philippa Campsie
Chinese Editor: Tommy Chung 鍾文昭
Production and Design: Justin Poy Graphics Inc.
Photograph of Author: Dr. Neville G. Poy

Printed and bound in Canada

First printing November 1995

Contents

This book is dedicated to all the descendants of Lee Hysan.

Acknowledgments

The writing of this book would not have been possible without the help of many Lee family members. The person I relied on the most is my fourth uncle, Dr. J.S. Lee, O.B.E., who not only gave me many of the books, photographs and reading materials for my research, but also gave me a lot of his time in relating to me events in the past. My cousins, Mr. Hon Chiu Lee and Mr. Peter Lee, have been most helpful in providing me with information, books, documents and photographs. My mother, Dr. Esther Lee, told me many stories of the past as did my aunts, Mrs. Ansie Lee Sperry and Mrs. Amy Lee Yuen. My aunt, Mrs. Ingrid Lee, was very enthusiastic in helping me obtain old photographs. I wish to express to them my greatest gratitude.

I also wish to extend my thanks to my late grand-uncle, Mr. Lee Shiu Yuen, my uncles, Mr. Lee Pak Chao, Mr. Davis Lee and Mr. Daniel Wong and my aunts, Mrs. Helen Wong, Mrs. Pearl Wong Kwong, Mrs. Josephine Wong Chiu and Mrs. Jennie Wong Hoo.

Also my heartfelt thanks to my long-time assistant, Peggy Ku, who helped me in the Chinese translation, not only of this book, but also in the Chinese research materials. I would also like to extend my thanks to the Chinese scholars who contributed to this book, Mr. Ku Han Chao, Mr. Dai Kwok Choi and Rev. David C.P. Lew, and to the very helpful staff of the Hong Kong University library and the Hong Kong City Hall library.

Last but not least, I am very grateful to my husband, Dr. Neville Poy, for taking all the photographs and reproducing all the old photographs in this book. My thanks to our son Ashley who gave me invaluable reference books, and to our sons Justin and Carter, and to Miss Conwell Tam, for their constructive criticism, and to our daughter-in-law Barbara, who spent many late nights helping Justin in the publication of this book.

A RIVER NAMED LEE

Prologue: Assassination

Hong Kong, April 30, 1928. At one o'clock in the afternoon, a dozen members of the Yue Kee Chinese club on Wellington Street were having tiffin, served by *foki* (waiters). Suddenly, they were startled by the sound of three shots fired in quick succession, followed by cries of "Save Life." Two club *foki* rushed down the staircase.

One *foki* by the name of Law Lau was the first to reach the bottom of the stairs. Going along the cement corridor, he saw the victim holding on to the wall. His face was extremely pale. It was obvious that he was seriously injured. Law Lau, however, paid little attention to him, but followed a man in white trousers and a short jacket, who darted through the passage leading from the club. By the time Law Lau had turned the corner, however, the man was nowhere to be seen.

By the time other members of the club found the victim lying in the corridor, he was dead. He had been shot three times. One bullet had pierced the left shoulder, another had lodged in the stomach and a third had entered the back of his neck. The last shot had apparently been fired when he turned around. One bullet had struck the cement corridor after striking the victim.

Two Indian constables arrived on the scene shortly afterwards. An ambulance was called, and the body was taken to the mortuary. Then the Hon. Mr. E.D.C. Wolfe, Captain Superintendent of Police, arrived with Detective Inspector Murphy, Sub. Inspector Lane and other officers. The family of the victim was called and the wife, the concubines, and a number of friends came to identify the body.

A revolver was found in the stairway of the Yue Kee club. It had been thrown away by the murderer when he made his escape, and was found to contain three empty chambers and three live shells.

The *South China Morning Post* of Tuesday, May 1, 1928, called the assassination one of the most sensational crimes ever perpetrated in the colony. The victim had been one of the best-known businessmen in Hong Kong, and the crime had been committed in broad daylight in the centre of the city. It was clear that the murder had been well planned. Club members confirmed that in addition to regularly taking tiffin at the club, the victim had frequently eaten his evening meal there as well. His custom of visiting the club at tiffin time was well known. The assassin was able to choose a quiet spot, and knew with a degree of certainty what time his victim would be going into the premises. One member of the club stated that about five or six years before, a rich Chinese merchant named Lo Chuk Wan had been assassinated on the steps, but this crime had been committed in the cement corridor outside the main entrance to the club. The murderer had apparently concealed himself at one side of the corridor and fired at his victim from behind.

The club was situated on the third floor of a building on Wellington Street, but the main entrance was on Kau Yu Fong Lane. This was a quiet thoroughfare, generally deserted by pedestrians during the greater part of the day. Joining Wellington Street and Kau Yu Fong Lane was a flight of stone steps, leading to a small lane that gave direct access to the club.

Kau Yu Fong Lane had once been a somewhat notorious street, the rendezvous of undesirable characters, especially members of alleged secret societies. However, as more business premises had opened in the immediate neighbourhood, the street had become a quiet backwater, and little frequented.

The motive of the crime was unknown, although the police discovered that the victim had received threatening letters. One of these warned that bombs would be thrown at the wedding of his eldest son, which had taken place shortly before the murder, although the threats were not carried out. Intimate friends had advised him not to visit establishments in lonely places. The fact that he had ignored these warnings had cost him his life.[1]

Prince of Zhau

As a little girl, whenever I bowed to my ancestors in our ancestral hall in the Big House, I was fascinated by the plaque on the altar that related the origin of our family. I had thought that my family came from Hong Kong, but in fact we came from Honan province in north China. My ancestors moved first to Sunwui in south China, and then to Hong Kong.

In 1122 B.C. Chou displaced Shang as the ruling dynasty in China (1122-255 B.C.). Ji Fa became the Emperor and assumed the title of "Wu Wang," the Martial Emperor. His father Ji Chang was consecrated as "Wen Wang," the Cultural Emperor. Feudalism had reached its highest degree of development.

One of the Emperor Chou Wu Wang's teachers was a man by the name of Chou Hung, my remote ancestor. During the reign of the Emperor Chou Sing Wang (1115-1078 B.C.), the Emperor wanted to reward the descendants of the teacher Chou Hung. Chou Hung's great-grandson, Hung Yik, was granted the feudal kingdom of Zhau at Ging Zhau, which is the southwest section of present-day Szechwan province. Lai Zhen, son of Hung Yik, the prince of the Kingdom of Zhau, was later assigned the area along the Lee River in Honan province as his feudatory, and Lee was designated by the Emperor as Lai Zhen's family name.

Throughout our family's history, our Lee ancestors have contributed to and glorified the times in which they lived. During the Ch'in dynasty (221-206 B.C.), general Lee Gai helped Shan Yu, Emperor of Sai Zhau, who, with his accomplice, the commoner Liu Bang, put an end to the Ch'in dynasty of Shi Huang Di, the first emperor of China. Liu Bang then defeated Shan Yu and became the first emperor of the Han dynasty. Shan Yu committed suicide and general Lee Gai was conscripted into Liu Bang's army. However, Lee Gai could not remain loyal to the new emperor, and was subsequently defeated by him. The emperor nevertheless pardoned him and gave him a piece of land to till for the rest of his life.

During the Han dynasty (206 B.C.-221 A.D.), Lee Chien was mentioned in many historical documents as the highest minister to the Emperor Chung San, the grandson of emperor Liu Bang. Lee Chun Lum was a high court official, the equivalent of a prime minister. Lee Zhen Yuan was a famous religious leader in Eastern Han (25-221 A.D.). During the dynasty of Eastern Tsin (317-420 A.D.) in south China, Lee Ro was the secretary of state.

During the T'ang dynasty (618-907 A.D.), there were 198 surnames in China. Lee was one of the five surnames in Yiunan and one of the five surnames in Honan (Yuinan and Honan form the present-day Honan Province). Lee Kai gained distinction as a great literary figure by being the calligrapher of the well-known inscription in the T'ang pagoda, Yan Ta.

Lee Shen was a well-known intellectual and philosopher in the Sung dynasty (960-1279 A.D.). During the Ming dynasty (1368-1643 A.D.), Lee Ben Chen contributed greatly to social and national security, and was instrumental in stopping invaders on the Chinese border.

During the Ch'ing dynasty (1644-1911 A.D.), the twenty-third year of the emperor Tao Kuang (1843), Lee Yi Quan sat for the Imperial examination and passed as Jing Shi (a successful candidate of the national civil service examination held at the imperial capital). He was appointed high minister and the event was described on a stone tablet in an important temple.[2]

From the time of Lai Zhen to the end of the Sung dynasty, over a period of almost 2500 years, the Lee family expanded from Lee River in Honan province to Nam Hung in Kwangtung province.

In the Southern Sung period (1127-1279 A.D.) near the end of the Sung Dynasty, during the lifetime of Lee Ken Long, son of Teh Chao, there was a mass migration during which the entire population of Nam Hung Fu in Honan province in north China moved to Kwang Chou Fu in Kwangtung province in south China. In our ancestral record, there are three different versions of the reason for this mass migration.

According to the first version, the Emperor Whe (1265-1275 A.D.) in Southern Sung had a favourite consort, Lady So. Lady So was very beautiful but had an unpleasant disposition. This made the Emperor most unhappy and his health suffered. Lady So was therefore banished to a remote part of the palace, and was not allowed to communicate with the Emperor.

Lady So decided to flee, dressed as a traveller. As she was about to leave Hang Chou, the capital of Southern Sung, she met a wealthy merchant called Wong at the pier. Wong was from Nam Hung Fu and had come to the capital to deliver shipments of provisions. He was on his way home when he met the disguised Lady So. She approached his boat and spoke to him at length. She asked to go with him and be his wife. Impressed by her beauty, Wong took her home with him, not realizing that she was the Emperor's consort. She had told him her name was Cheung.

Time passed. The Emperor decided to recall Lady So from banishment. When he discovered that she had escaped, he was very angry. He sent messengers to look for her, but they were not able to find her. Meanwhile, in Wong's household, one of the servants had discovered that the wife the master had brought back from the capital was none other than the Emperor's consort. The secret was out.

When the Emperor heard the news, there was great fear in Nam Hung Fu. It was decided that for the Emperor to save face, an excuse would have to be made that there were bandits there, and soldiers would be sent to kill the entire population.

One family with the surname Law, in Nam Hung Fu, had a son-in-law called Leung, who was an official in the capital. Leung heard of the impending disaster and immediately sent a messenger to his in-laws to tell them that they must leave Nam Hung Fu immediately. Ninety-seven families from fifty-eight villages got together to apply for permission from the authorities to leave, saying that it was too difficult to make a living in Nam Hung Fu, and therefore they wished to move further south where the land was more fertile. They submitted their application on the tenth day of the first moon. On the fifteenth day of the fourth moon, in the ninth year of Hum Shun in Southern Sung (1274 A.D.), the entire village migrated to Kwang Chou Fu, with the exception of Lee Ken Long's family and Leung's family. Because the heads of these two families were officials in the central government, they were not in any danger. Ken Long was a scholar who held a position equivalent to that of a senator in the capital.

The second version of the migration story concerns a very powerful high court official named Gar Si Doh, who lived in Southern Sung. One day he was to accompany the Emperor to the temple to pray. General Wu asked the Emperor which horse he would like to ride, and he said, Siu Yin Zi. Gar Si Doh was angry with General Wu because he had not been asked first. He pretended to be ill and excused himself from accompanying the Emperor. When the Emperor heard that Gar Si Doh was upset, he dismissed General Wu. General Wu's sister Lady Wu, who was one of the Emperor's consorts and mother to his son, was sent to a nunnery.

The years went by and the Empress had no son. The Emperor decided to move his son by Lady Wu to the east palace and make him crown prince. The court officials, however, advised the Emperor that the crown prince must have his mother near him, therefore Lady Wu must be recalled from the nunnery. At this point, it was discovered that Lady Wu had not gone to the nunnery but had sought refuge with the villagers of Nam Hung Fu. When the villagers learned that the Emperor wanted to recall Lady Wu to court, they were afraid and gathered together to make application to leave in the ninth year of Hum Shun.

The third version of the story also concerns Gar Si Doh, but the details vary. In the eighth month of the ninth year of Hum Shun in Southern Sung, the Emperor chose a date to pray in a famous temple. He asked the education minister, Gar Si Doh, to accompany him. When General Wu Yu Jo prepared the wagon, it was raining very hard. Gar Si Doh asked General Wu to wait until the rain had stopped before proceeding. General Wu, however, gave the Emperor the impression that Gar Si Doh did not want to accompany him. The Emperor went to pray alone. Minister Gar took this as a slight to his position and resigned. The Emperor was distressed by the misunderstanding caused by General Wu and had him imprisoned. General Wu's sister, Lady Wu, who was one of the Emperor's consorts, tried to save her brother, but her efforts resulted in her own banishment from court.

When she left the court, Lady Wu met a prosperous businessman who brought her to his home in Chu Gai Lai in Nam Hung Fu. However, when he discovered who she was, he refused to allow her to spend the night. Having no place to stay, Lady Wu fell into the hands of strangers and was molested. Alone and friendless, she drowned herself in a river. When the Emperor heard about her suicide, the villagers in Nam Hung Fu were afraid. They therefore banded together to make application to leave.

Nine years later, in 1283 A.D., after Kublai Khan succeeded in conquering the entire country, the Sung Dynasty was replaced by the Yuan Dynasty and there was unrest all over the land. Lee Ken Long decided to move his family south to Kwang Chou Fu to rejoin the villagers, and settled in Sai Far Kue in Sunwui city. The family lived there for ten years and later moved to Sunning.

When he died, Lee Ken Long was buried in Sunning in a very good plot of land. His descendants were not prosperous, however, and the burial plot was sold to a family by the name of Yu. Later, the family was able to buy back the burial plot because it was considered to be lucky. The importance of the burial site is repeatedly mentioned in our ancestral book.[3]

The Lee River in Honan province has since dried up and no longer appears on the map of China.

Golden Mountain

During the Ch'ing dynasty (1644-1911), in the twenty-second year of the reign of Tao Kuang, 1841, on the twenty-second day of the third moon, my great-grandfather Lee Leung Yik was born into a humble family in the village of Hoiping in Kwangtung province in southern China. Great-grandmother was born in 1850, on the twenty-sixth day of the sixth moon. Together they had four sons and two daughters. The family lived in Sui Jiang Hang, in the village of Hoiping, which consisted of twenty to thirty families.

The life of the villagers was hard. The houses were small, dark and dingy with earthen floors. The villagers worked from dawn to dusk, growing rice, vegetables, oranges, tangerines and lichees, and raising chickens, ducks and pigs. The land was not fertile and the villagers barely had enough food for their families.

Life for the women from poor families was particularly harsh. They had to do all the cooking, mending and washing, raise the children, and work in the fields as well. Giving birth was simply an interruption in a regular day's work for these women. They would run home when labour started if they had the time; otherwise, the delivery would be in the field. The mother often delivered her own baby. If a knife was not at hand, a piece of pottery would be smashed to give a clean sharp edge to cut the baby's umbilical cord. The baby would then be wrapped in a piece of cloth and the mother would go right back to work.[4] People survived if there were no external calamities.

In the latter part of the nineteenth century, the Chinese population suffered greatly under the ineffective rule of the Ch'ing dynasty and the tyranny of the ruthless warlords. The country was impoverished by the constant bullying of European countries trying to get trade concessions from China. There were droughts, floods and famine. The people in south

China were the hardest hit by the political and economic hardships. Many men emigrated to work as indentured labourers under contract, commonly known as the "pig trade." Many went to the East Indies and Malaya; others went to the West Indies and islands in the Indian Ocean; still others to Australia. They were packed into crowded and insanitary boats known as "floating hell." There were often riots that killed the captain and the crew. Many died of hunger or disease. The highest death rate recorded of this kind of human cargo was as high as 66 percent.[5] When they reached their destinations, they worked under harsh and restrictive conditions, living in compounds and often subjected to corporal punishment.

Chinese emigration was actually illegal, but it had been going on for a long time. This overseas emigration received great impetus from the discovery of gold in California in 1849 and in Australia in 1851.

In 1890, Great-grandfather Lee Leung Yik journeyed from Kwangtung province to San Francisco, in the hope of making some money to send home to help support his family. San Francisco was known as Gum San, the Golden Mountain. Despite his age (he was forty-nine), I believe Great-grandfather went with the hope of panning for gold. But if he had tried, he would not have been successful, because Chinese miners were barred from the rich strikes until white miners had exhausted the find. Most of the original Chinese immigrants ended up working as common labourers, ranch hands, cooks or at whatever jobs they could find. Some became disillusioned and returned to China. Great-grandfather worked at the Union Cigar Co. at 17 Dupont Street. The company was listed in the 1876 *Chinese Business Directory* in San Francisco.

After having settled in San Francisco, Great-grandfather sent for his younger brother, Man Yik, and for his second son, Ting Sien, my grandfather. Man Yik also brought his eldest son with him.

My grandfather, Lee Ting Sien, later known as Lee Hysan, was born on the twentieth day of the ninth moon in 1880, in the village of Hoiping in Kwangtung province. He was nine or ten when he travelled to San Francisco to join his father. Why Great-grandfather sent for his second son and not his eldest was anybody's guess, but one answer was that the eldest had to stay home to help his mother. Or perhaps Great-grandfather was already able to see the potential in his second son.

The Chinese community in San Francisco in the early days tended to be self-contained. The Chinese were regarded as a curiosity. They dressed in cotton tops and pants that looked like pajamas, with cotton shoes, and they wore their hair in long queues. The style had started in the seventeenth century when the Manchus conquered China, and Ch'ing took over Ming as the ruling dynasty. The Manchus wanted to degrade the Han Chinese to make them look like horses, therefore the men had to wear their hair in long queues to look like horses' tails, and their shoes looked like horses' hooves.

Since most of these Chinese immigrants to America planned to return home when they had enough savings put aside, their ties with their families and villages in China were very strong. They depended upon their families and district associations for guidance and assistance in the New World. These associations, which existed in some form almost from the time the first Chinese immigrants reached the shores of California, helped establish Chinese language schools and provided leadership in the long battle against discrimination. The traditional Chinese customs, festivals and operas helped to maintain the morale of the early immigrants. Disputes were usually handled either by the district association or the Chinese Consolidated Benevolent Association. The tongs, rather than the police, often provided protection.

The original Chinese immigrants to San Francisco were all from Kwangtung province in southern China. The Chinese Consolidated Benevolent Association, popularly known as the Chinese Six Companies, was the watchdog that oversaw the affairs of the Chinese community. The first association was established in 1849 and known as the Kwang Chou Association because it included immigrants from six districts of Kwang Chou in Kwangtung province. In 1850, the Sam Yup Association was formed for immigrants from the three districts of Namhoi, Punyu and Shuntak. Before the end of the same year, the Sze Yup Association was formed for immigrants from Yanping, Hoiping, Sunning and Sunwui districts. The fourth and fifth associations were established in 1852, the Young Wo Association for immigrants from the Heungshan area, and the Hip Kat Association for the Hakka compatriots. And in 1854, the sixth organization, the Ning Yung Association was formed.[6] Great-grandfather and his brother belonged to Sze Yup Association. Since all Chinese immigrants came from small villages, they all spoke village dialects. Anyone who spoke the same dialect was regarded as a brother.

During the years when the Lees lived in San Francisco, Ting Sien and his cousin attended the Chinese Primary School at 929 Clay Street.[7] This school was originally known as the Chinese School in 1859 and was established by the Board of Supervisors to ensure the segregation of Chinese children from white children in San Francisco's public schools. It was later renamed the Oriental School in 1915, and in 1924 the name was changed to the Commodore Stockton School.[8]

The churches provided the first and, for many years, the only English-language classes for residents in Chinatown. The English that Grandfather learned in San Francisco became a great asset to him in later years.

In 1896, after having saved some money working in San Francisco, Great-grandfather, together with his brother and the two boys, returned to China.

Back in Hoiping, Great-grandmother and her family had found life very difficult. However, they had noticed that in the nearby area of Sunwui, the land around Gyan Chou Wu (Silver Lake) was regularly flooded by the river. It was not only a bountiful area for hunting, but the land was also very fertile.[9] When Great-grandfather and Ting Sien returned to China, the family moved to Sunwui and lived in the village of Garlieu in Sheungsui County. From then on, Sunwui became our ancestral home.

Great-grandfather was regarded as affluent since he came back from Gum San with some savings. A sizable piece of land was purchased for the building of a new home in Garlieu village for my Great-grandparents' retirement and a new ancestral hall was built.[10] Unfortunately, the original home was demolished during the Cultural Revolution, and all the bricks were removed to be used for building other structures.[11] The present building on the same site is modern, with a fair-sized garden surrounding the house and a number of fruit trees. The house is occasionally used by Uncle Lee Pak Chao, a relative, whenever he is in Garlieu.

A Little Island

Great-grandfather was content to retire to his village in China, but it was not the kind of life he wanted for his children. In 1897, he had his second son, Ting Sien, my grandfather, enrolled in Queen's College in Hong Kong. Great-grandfather would have known about the colony because of his travels. When he had gone to San Francisco, he would probably have boarded a boat in Hong Kong, because Hong Kong was the main port for the transportation of Chinese abroad. When he sent money home to China, he could not send it directly, and it would have been sent via banks in Hong Kong. He would have heard of the great success of a number of Chinese merchants in the British colony.

Hong Kong, well known for its magnificent harbour, was first mentioned in Chinese historical records in the Ming Dynasty as the name of a small village on the south coast of the island. The island and its surrounding areas were famous for the growing of incense trees which produced incense wood known as Kwun Heung. During the Ming Dynasty, incense wood and incense products were assembled at Shek Pai Wan (Aberdeen Harbour) before being shipped to Canton and as far north as the provinces of Kiangsu and Chekiang. This area was therefore called Heung Kong, meaning fragrant harbour, and the village was called Heung Kong Tsuen, meaning fragrant harbour village.

During the Ming Dynasty (1368-1643 A.D.), economic conditions improved and the population increased. During the Wan Li reign at the end of the sixteenth century, there were at least seven villages on the island, including Heung Kong village. These were mainly on the south coast; the north coast was sparsely populated. At the end of the Ming Dynasty, the island was frequently attacked by pirates, despite the frequent patrols of naval vessels along the coast.

During the early part of the Ch'ing Dynasty, in the first year of the Kang Hsi reign in 1661, coastal evacuation was carried out: the population on the island was moved inland and the coastal villages were abandoned.

Eight years later, the Edict of the Coastal Evacuation was revoked. The population returned and rebuilt their villages. In the early years of the Yung Cheng reign, the seven villages of Hong Kong, Tai Hang, Chung Hom, Chek Chue, Tai Tam, Shau Kei Wan and Wong Nai Chung were rebuilt. The government constructed forts and set up military posts along the coast to discourage piracy, but none was established on the island.

During the reign of Chia Ch'ing (1796-1820), two more villages were built, as well as the Hung Heung Lo Naval base, which was under the control of the Tai Pang Battalion.[12] In 1841, the sparsely populated, rocky island was ceded to Britain.

This little island became home to the Lee family at the end of the nineteenth century. Its status as a British colony was the reason for their move, since if the island had remained part of China, conditions there would have been the same as in south China. In order to relate how the island became a British colony, I must go back in history.

The Opium Question

Throughout the seventeenth and eighteenth centuries, goods such as tea, china and silk were very much in demand by the West, and trade between China and Britain, France, Holland, Portugal and the United States increased. Trade was, however, greatly limited by the restrictions imposed by the Imperial Chinese government. The government maintained that the Celestial Kingdom produced everything the country needed and therefore had no need for manufactured goods from the "foreign devils." Foreign merchants were required to pay for goods in silver.

By the end of the eighteenth century, Britain was the leading country trading with China, mainly because of the popularity of tea in England. In 1664, slightly more than two pounds of tea was imported into England. By 1783, the legally imported amount was 2,600 tons. Probably three times that amount was smuggled in to avoid high duties.[13]

Trade was limited to Canton, which acted as an outlet for Chinese goods; the other ports were closed to foreigners. Aside from tea, silks, cottons, porcelain, paper, medical products and spices were also in demand, and payments were strictly cash. Because of the winds, the trading season was limited to the months from May to November. During those months, the foreigners were allowed to reside, not in the city of Canton, but in their "factories" outside the city. They kept their permanent establishments downriver at Portuguese Macao.[14] They were not allowed to learn Chinese, to employ servants, to use sedan chairs, to possess firearms, or to have their wives with them. However, these restrictions were not always enforced.

Trade with the foreigners was strictly regulated and under the authority of the Hoppo, the superintendent of the South Sea Customs. The Hoppo held a medium rank far below that of the Viceroy of Kwangtung and Kwangsi, or even that of the Governor of Canton.

The latter gentlemen were scholar-administrators who received modest salaries and whose lifestyle depended on "gifts" from the Hoppo. In theory, Chinese customs duties were clearly specified: in practice as much as possible was collected, and as little as prudent was remitted to Peking.

Under the Hoppo was the merchants' guild, or Kung-hang (Cohong) which in turn, was "squeezed" by the Hoppo. Foreigners were allowed to deal only with the Cohong merchants. It was necessary for the Cohong merchants to take every opportunity to try to secure large profits from the foreign merchants. This severely limited the foreign merchants' ability to bargain and get the best prices for their goods.

Officials from the various embassies were sent to Peking to try to negotiate better relations between their merchants and the Chinese authorities. These attempts failed because the Imperial Court refused to recognize that any other country could be regarded as the equal of China.

Despite the restrictions, trade between China and the West increased. Western governments were concerned about the drain of silver from their countries. The merchants were eager to find some other commodity that the Chinese would buy from them. They found opium.

Opium was first grown in China for medicinal uses. In the seventeenth century, the habit of smoking it had gradually spread. To control opium addiction, a government monopoly had been established in 1729, and high taxes restricted the drug to a privileged few.[15] In 1800 the import of opium was forbidden. However, British merchants saw a huge potential market among the poorer people of China for cheap Indian opium. Western traders in Canton and Macao smuggled in massive quantities of Indian opium with the cooperation of Chinese merchants and officials who received "gifts" as payment. This trade caused a huge increase in addiction, including the eunuchs in the imperial household. Millions of pounds of opium were imported or smuggled into the country. All the tea in China could not pay for it. Chinese silver reserves evaporated. The Chinese government passed edicts forbidding the smoking of opium, but these proved to be useless.

As Britain was approaching the zenith of its power in the middle of the nineteenth century, China had sunk to its lowest point since the Manchu took control in 1644. Corruption was endemic. The struggle over the opium trade represented China's fight for sovereignty, even though corrupt Chinese and Manchu officials participated in the trade. Westerners provoked the Chinese at every turn, and when the Chinese struck back, the foreign merchants demanded concessions from local mandarins. If concessions were not forthcoming, gunboats were called in. No foreigner seriously believed that the Manchu had the strength of character to take drastic action.

By 1835 the opium imported into China had risen to over 30,000 chests. The increase was alarming. Since all opium was imported illegally, no duties were paid, except for exactions going to the middlemen and the mandarins' private accounts. And since all illegal imports had to be paid for in bullion, the strain on China's reserves was great. In 1836, the Peking government debated whether to suppress or to legalize the trade. The Canton merchants were unanimous in asking that the trade be legalized. It was reported that the Empress supported it. Senior Judge Hsu Nai-tsi also wrote a paper on this subject which was translated and published in the *Canton Register* of July 12, 1836.[16]

Both Europeans and Chinese in Canton took legalization for granted. James Matheson of the British merchant company Jardine Matheson and Co. disagreed, however. He wrote on July 12: "I do not think well of the plan as far as our interests are concerned — tho' it has already enhanced prices."[17] Six months later, nothing had happened. It soon became clear in Canton that, far from legalizing the drug, Emperor Tao Kuang had made his decision that the drug was to be outlawed, and the only question was how best to put an end to the trade.

In the early 1800s, there were two main British firms trading in China: Dent and Co. and Magniac and Co. The latter became Jardine Matheson and Co., the partners being Dr. William Jardine and James Matheson. These British firms were colloquially known as hongs, and their partners as taipans; Jardine Matheson was Ewo Hong, and Dent's was known as Pao-shun Hong.

James Matheson and William Jardine went to England in 1836 to encourage the British government to take a stronger approach towards China. They suggested that if the Chinese refused the demands of the merchants, then England should consider taking over a territory in China such as

Chousan, an island off central China. Hong Kong was far from people's minds at the time.

In his book *The Present Position and Prospects of the British Trade with China,* Matheson described the Chinese as a people characterized by a marvellous degree of imbecility, avarice, conceit and obstinacy who subjected innocent foreigners to injuries and insults of horrible description. He insisted that Britain must vindicate its insulted honour as a nation and protect the interest of its commerce.[18]

Charles Elliot, who was the Superintendent at the time, complained to the Foreign Secretary, Lord Palmerston, that the British merchants' behaviour was the chief cause of offence, since they treated the Chinese insultingly. Unfortunately, Elliot was not empowered to remove or punish anyone. For its part, the East India Company maintained that the Chinese were a great, powerful and peculiar people who would never be bound by any treaty, and in the absence of any treaty, the law of nations prohibited any attempt to enforce British claims upon them.[19]

The American traders saw the writing on the wall. Russell's, the largest American trading house, decided to discontinue all connection with the opium trade in China, a business that was fast becoming too dangerous and disreputable.[20]

The British traders, with more at stake, expected that if the worst came to the worst, they would be bailed out by the Royal Navy. However, as a precaution, Jardine Matheson suggested that deliveries of opium and piece goods should be diverted to Hong Kong and up the coast. This indicated that some trade was already taking place in the island's waters.[21]

In March 1839, Lin Tse-hsu, Governor of Kiangsu, was appointed Imperial Commissioner with full powers to deal with the opium question in Canton. As viceroy of Hupeh and Hunan provinces, Lin had ordered that any smoker who refused the cure would have part of his upper lip cut off, to prevent him from using a pipe.

Lin made it clear that his targets were the Chinese collaborators. As long as the foreigners obeyed the law and refrained from smuggling, they would be benevolently treated. He knew that legitimate trade was important, and must be protected. On March 18, he proclaimed that all opium stocks must be surrendered and that foreigners must pledge never again to deal in the drug. If these conditions were not met, the Cohong merchants would suffer imprisonment, expropriation and decapitation. The foreigners would suffer severe punishment prescribed by the new law. Three days were given for compliance, and in the meantime, the foreigners were to be confined to their factories.[22]

On March 21, at a meeting of the Chamber of Commerce, the foreign merchants agreed to surrender just over a thousand chests, a very small quantity. Lin was furious at what he recognized as blatant procrastination. Since William Jardine was still in England, Lin ordered the arrest of Lancelot Dent of Dent and Co., who was identified as the chief of the foreign smugglers.[23] Ironically, Dent was more popular than Jardine with the Chinese. Captain Elliot tried to smuggle Dent out of the country, but the Chinese blocked their escape by constructing barriers across the river. Dent was subsequently freed. He later bought a house in Manila to use as a possible alternative centre for his opium distribution.

Captain Elliot, on behalf of the British government, instructed the merchants to surrender the drug, which in essence meant that the British government was obliged to compensate them. The government had no intention of paying unless it could cajole or coerce the Chinese into reimbursing it. More than 2.6 million pounds of opium was delivered and burned at a specially constructed site by the banks of the Pearl River.[24]

Jardine Matheson agreed not to import any more opium. James Matheson knew that he could argue later that this agreement had been made under duress, and was therefore not binding. He wrote to his suppliers and told them to send their opium to Alexander Matheson in Macao, who would be happy to continue to dispose of it.[25]

Foreign secretary Lord Palmerston asked for the advice of the London East India and China Association, which they gave in a long letter of November 2, 1839. The Association was ready to accept that British subjects in China would be subject to Chinese laws, including the ban on importing opium. "We are quite prepared to admit, should the Chinese persist in prohibiting the import of opium, that henceforth the British merchants trading to China must obey the laws of that country in respect to that article, and that the Crown of Great Britain cannot be called upon to interfere in any manner in support of its subjects who violate them." They also considered that it was essential that British representatives be allowed direct access to the Chinese authorities, upon equal terms, and they specified ports in addition to Canton that should be opened for trade. If this was not allowed, then they recommended "the cession, by purchase or otherwise, of an island be obtained." The suggestions were adopted in their entirety by Lord Palmerston.[26]

The memorandum was supplemented by a delegation from the Association led by John Abel Smith, Whig M.P. and banker for Jardine Matheson, William Jardine, shortly to become Whig member for Ashburton, and Alexander Matheson. The deputation could not persuade Palmerston to yield an inch on the question of the government's paying, then and there, for the surrendered opium. They did, however, succeed in getting Palmerston to accept that the Chinese could be forced, without too much difficulty, to pay up instead. The delegation provided practical advice on how this should be done. Palmerston had little difficulty in convincing the rest of the cabinet of the need to dispatch an expedition to China.[27]

Commissioner Lin ordered all foreign traders to sign personal bonds of life and death before their ships were allowed back to Whampoa. (Whampoa, about seven miles below Canton, was the furthest inland that ocean-going ships could go up the Pearl River because of the sandbar.) Danish, German, American and Spanish captains signed Commissioner Lin's pledge and entered Whampoa to trade freely in other goods, but Captain Elliot, in a rage, tore the document to shreds. Henceforth, he announced, all British vessels would discharge their cargoes at Macao.[28] When Commissioner Lin prohibited commerce by the British at Macao, the British merchants traded directly with Chinese smugglers offshore.

British Colony

On July 7, 1839, a drunken party of sailors got into a fight at Tsim Sha Tsui, Kowloon. A British sailor killed a Chinese civilian, Lin Wei-hsi, and injured several others.[29] Captain Elliot paid hush money to the injured victims and compensation to Lin's family in return for a statement that his death had been an accident and staged a mock trial that found the killer innocent. When Commissioner Lin Tse-hsu demanded the surrender of the culprit, Captain Elliot refused. The sailor and his mates were shipped back to England and set free.[30]

On August 15, with Portuguese cooperation, all British residents in Macao were forced out of their homes by the Chinese government. They boarded sixty or so merchant ships, sailed for Hong Kong, and anchored there. Imperial edicts were sent to the villagers in Kowloon stating that fresh food and water were to be denied the British refugees. Captain Elliot was able to persuade the authorities to relent. On September 4, some provision boats owned by the villagers were loaded and set off towards the British, only to be intercepted by Chinese war-junks. Elliot opened fire from the *Louisa*. These first shots eventually turned into a war.[31]

How long Commissioner Lin would allow the British to stay anchored in Hong Kong harbour was anybody's guess. He intended to have them back in Canton by September to resume normal trading, provided that they accepted his bond and ceased trading in opium. Captain Elliot had forbidden British ships to accept the bond, and Commissioner Lin threatened to destroy all the ships. On October 26, after receiving Lin's ultimatum, Elliot warned all British ships to leave Hong Kong. After months of negotiations, on November 3, 1839, the *Volage* opened fire and sank four war-junks. The war was on.[32]

In London, William Gladstone, a member of the Tory opposition in the British parliament, denounced the government: "A war more unjust in its origin, a war more calculated to cover this country with permanent disgrace, I do not know and have not read of. The British flag is hoisted to protect an infamous traffic... we should recoil from its sight with horror."[33]

Commissioner Lin banned the British from Canton completely, with the result that the Americans, who had remained in the factories, simply took over the trade on the behalf of the British. Legitimate British imports went through as usual. Opium continued to be sold through ports up the coast. It was sent to Manila to be transshipped to the east coast of China. Jardine Matheson rapidly recovered the profits the company had lost through the confiscation of the drug.[34] Foreign secretary Lord Palmerston wrote to Captain Elliot that a naval force and a small army detachment would arrive at the end of March the following year. This force was to be under the command of Elliot's cousin, Admiral George Elliot.

Unfortunately, China misunderstood the special relationship that existed between private British traders and the government in London. Peking also underestimated the military power of England. The British Foreign Office had another motive for war: it wanted to force direct relations with the court in Peking, which the Manchu had consistently refused.[35]

Commissioner Lin, in the meantime, was replaced by an Imperial kinsman of high rank, Ch'i-shan, a hereditary Marquis, Governor of Chihli (present-day Hopeh province). His job was to soothe the barbarians and avert war.[36]

To China's surprise, the British did not pursue the attack on Canton, but sailed up the coast to capture Tinghai at the mouth of the Yangtze, where the defenders were unprepared. About 10,000 seasoned British troops were deployed against Chinese forces who had no experience fighting anyone but themselves and had no acquaintance with modern cannons. Other battles followed at Canton, Ningpo and other ports, until Peking gave in.[37]

On February 20, 1840, Palmerston wrote a letter to the minister of the Emperor of China instructing him that "The British Government demands that one or more sufficiently large properly situated Islands on the coast of China, to be fixed by the British plenipotentiaries [Admiral G. Elliot and

Captain C. Elliot], shall be permanently given up to the British Government." When the Admiral returned home, the choice was left entirely to Captain Elliot. The British government had been expecting to get the island of Chousan off the northern coast of China. This was not acceptable to the Chinese.

Lord Palmerston had also had in mind Ningpo as part of the settlement. William Jardine suggested Formosa (Taiwan), together with Quemoy and Amoy. Hong Kong, an island with a few fishing villages, was never mentioned. [38]

However, on January 15, 1841, Ch'i-shan wrote to Charles Elliot offering him either Hong Kong or Kowloon but not both. Elliot replied, accepting Hong Kong on January 16.

Captain Charles Elliot, being a career naval officer, preferred Hong Kong because of its magnificent harbour. Sir John Barrow, Secretary to the Admiralty, described Lei Yue Mun (at that time it was the name for the entire harbour) as excellent for ships of any size, and capable of being defended against a superior force in time of war.[39] The island also had an abundance of fresh water. With a few guns mounted, men to work them, and a warship, the island could afford protection to merchant shipping. Besides, the island of Formosa was larger than Ireland, and might be difficult to take and hold.

On January 28, 1841, at the Convention of Ch'uan-pi, Britain, represented by Captain Elliot, demanded the cession of Hong Kong to the British crown, but with a provision that duties should continue to be paid to the Chinese authorities. Arrangement was made for the payment of a $6 million indemnity (in Spanish silver dollars) in six equal instalments, direct diplomatic relations on a basis of equality, and the opening of the port of Canton.[40]

The British lost no time in making sure of Hong Kong. At 8 a.m. on Tuesday, January 26, 1841, the marines from the British naval force, commanded by Commodore Sir Gordon Bremer, landed on Possession Point at the foot of Taipingshan.[41] At 8:15 the Union Jack was hoisted in Pohingfong and the Queen's health was drunk with three cheers.

Two groups with the most immediate interest in the acquisition of Hong Kong were the merchants and the missionaries.[42] They understood the advantages that a piece of British territory in South China would afford them. Many merchants were present at the hoisting of the flag, including at least four Indian merchants. James Matheson of Jardine Matheson came from Macao to witness the event, and afterwards, he circumnavigated the island. On February 8, eight Protestant missionaries chartered a lorcha in Macao and went to Hong Kong on an exploratory outing. Their verdict was that: "Hong Kong will, if retained by the British, rise in importance and influence until it becomes the first insular emporium in these Eastern waters." [43]

Queen Victoria wrote to her Uncle Leopold, King of the Belgians: "Albert is so much amused at my having got the island of Hong Kong, and we think Victoria ought to be called Princess of Hong Kong in addition to Princess Royal."[44]

Captain Elliot, from on board H.M.S. *Wellesley,* proclaimed that Hong Kong was now part of Her Majesty's dominions, and that for the time being he himself would govern the island.[45] On February 1, proclamation was made to the Chinese inhabitants of the cession of the island of Hong Kong to the British Crown. However, hostilities continued.

By negotiating for the cession of Hong Kong, Charles Elliot had disregarded instructions sent to him from the government. In his letter to Elliot announcing Elliot's dismissal, dated April 21, 1841, Lord Palmerston was dismissive about Hong Kong: "You have obtained the Cession of Hong Kong, a barren island with hardly a house on it... Now it seems obvious that Hong Kong will not be a Mart of Trade... our Commercial Transactions will be carried on as heretofore at Canton; but they [the British residents] will be able to go and build Houses to retire to, in the desert island of Hong Kong."[46]

The arrangements at Ch'uan-pi were repudiated by the Chinese government. Ch'i-shan was packed off in chains to Peking and was replaced by Yang Fang.[47] On May 21, 1841, Chinese reinforcements were sent to recover Hong Kong. Fireships descended on the moored warships and guns from concealed batteries opened fire. However, H.M.S. *Nemesis,* with her mobility and fire-power, saved the situation for the British. A swift advance on Canton was ordered.[48]

Hostilities continued until July 1842. On August 29, 1842, the Treaty of Nanking was signed under duress aboard the H.M.S. *Cornwallis.* This was a turning point in the history of China. Five treaty ports were opened to foreign trade — Canton, Shanghai, Amoy, Foochow and Ningpo — $6 million in Spanish silver dollars was paid in indemnity for opium destroyed by Commissioner Lin, $3 million paid for debts owed to British traders by Chinese merchants, $12 million paid as indemnity to Britain for the cost of the war, and the island of Hong Kong was ceded to Britain. However, it was not until January 4, 1843, that the British government decided to keep Hong Kong, when Lord Aberdeen (who had replaced Palmerston as foreign secretary) asked Sir Henry Pottinger to assume the government of Hong Kong.[49]

On June 26, 1843, when the Treaty of Nanking was ratified, the Emperor of China finally recognized Hong Kong as a British colony.

Hong Kong in its early days was a disappointment. According to the census taken on May 15, 1841, there were sixteen villages on the island with a population of 7,450.[50] A deputation informed the colonial secretary on August 29, 1845: "Hong Kong has no trade at all and is the mere place of residence of the Government and its officers with a few British merchants and a very scanty poor population."[51]

On March 8, 1851, the *Economist* described Hong Kong as a hilly, stony island with an excellent harbour. The only successful merchant houses were Jardine Matheson and Dent, which were pursuing their customary trade. However, the article suggested that things might improve and that Hong Kong could become a useful settlement. It was, at any rate, a refuge for the China trade.[52]

The proximity of Hong Kong to the mainland worried the British naval and military men from the earliest days. Forts on Kowloon held the island within easy cannon shot. Although most of the peninsula was occupied by insignificant hamlets tenanted by stonecutters and lime burners, Kowloon City, at the northeastern corner of the Kowloon peninsula, had been fortified in 1668 when a signal station was established.[53]

The Chinese were well aware of Kowloon's strategic position. In 1843, after the ratification of the Treaty of Nanking, the Imperial Chinese government transferred the assistant magistrate of Hsin-an county to Kowloon, with administrative responsibility for 491 villages. Also transferred there was the Commodore of Ta-p'eng, the chief military officer of the county. The garrison was increased to 150.[54]

It soon became apparent that these measures were not enough. In 1846, Ch'i-ying, the governor-general of Kwangtung and Kwangsi, sent a memorandum to the Imperial court, pointing out the exposed position of the site, and suggesting that a walled city be constructed and equipped with cannon. He also proposed building offices and barracks, not only to provide accommodations for the civil and military personnel (who had hitherto been billeted in private homes), but also to create facilities for drilling. Such measures, he felt, would have a "constraining effect" on the barbarian base in Hong Kong, and would greatly strengthen the coastal defence of the area. Kowloon Walled City was completed in 1847.[55] The Hsin-an magistrate, commenting on the founding of the Lung-chin Communal School, declared that since Kowloon had become a point of interaction with barbarians, the inhabitants needed to be fortified morally, thus making a school necessary. He even hoped that the inhabitants of Kowloon might exert a civilizing influence on the intruders.[56]

The Chinese officials in Kowloon and the British officials in Hong Kong kept in close touch and generally cooperated in maintaining law and order in the vicinity.

At the Convention of Peking in October 1860, Kowloon peninsula was ceded to Britain. However, the Kowloon Walled City remained part of China. The Chinese officials there remained actively involved with the community on both sides of the border, as numerous temple inscriptions in Hong Kong testify.[57]

The Chinese community in British Hong Kong was obviously well aware of the Chinese official presence across the harbour in the Walled City. Sometimes they looked to it for protection. For instance, in 1886, when it was rumoured that 500 children would be sacrificed to consecrate the Tai Tam Water Works, children were sent to Kowloon City for protection, to the extent that hardly any children were to be seen in Hong Kong for two days.[58]

On July 1, 1898, the New Territory (as the New Territories were then called), the area south of the Shumchun river, was leased to Britain for ninety-nine years. The population in this area was tradition-bound and indignant that their homeland had been transferred to the British without their being consulted. The day before the hoisting of the flag, the police and the guard of honour discovered a trench on the side of the hill, about 800 yards from the spot where the flagstaff was to be erected, filled with antique weapons to ambush the officials and the ladies who were to be present at the ceremony. Soldiers were sent to chase the peasants over the hills. They managed to capture enough ancient weapons to fill an armoury. As a result, the picnic arrangements at the hoisting of the flag were cancelled.[59]

In 1899, the Chinese officials left the Walled City in Kowloon.[60] Hong Kong as we know it today consists of Hong Kong Island, Kowloon Peninsula, the New Territories and 235 adjacent islands.[61]

In 1889, Rudyard Kipling asked in Hong Kong: "How is it that everybody here smells of money?" and was told: "It is because the island is going ahead mightily. Because everything pays." That summarized Hong Kong as it was then and remains today.[62]

The Pupil Teacher

Ever since Hong Kong was ceded to Britain, it had become a refuge for the Chinese from China, despite rampant discrimination. At first, the outlaws and the desperate came. But during the Taiping upheaval (1851-65) in China, Hong Kong came to be regarded as a desirable place to live by "respectable" Chinese. It was a free port, and there was no war. Many Chinese merchants brought their families to live in Hong Kong.

These were people of substance and enterprise. Businesses flourished. By 1855, the British government no longer had to subsidize Hong Kong. Hong Kong became the main trading port for south China, handling one quarter of China's imports and one third of its exports.[63] By 1859, sixty-five Chinese firms were large enough to be registered as hongs.

From 1861 onwards, Hong Kong became the economic centre for trade and manufacturing, with sugar refineries, cement factories, weaving and paper making. With the establishment of banks and insurance companies and the availability of commercial loans, smaller companies were able to compete with the large establishments. Despite the fact that the Chinese were treated as second-class citizens, they were commercially very successful.

The largest and the most powerful hong at the time was Jardine Matheson and Co. Its officers were very influential in the British Parliament, and effectively ruled Hong Kong from the governor down. They sat on the Legislative and Executive Councils, they controlled the police, and were members of the exclusive Hong Kong Club and the Jockey Club.[64]

In 1866, the financial world was rocked by the collapse of the London discount house Overend and Gurney. On the China coast Jardine Matheson avoided bankruptcy by negotiating a capital sale, and lease-back of their extensive property portfolio, but Dent and Co. had to shut its doors. Fifteen years later, only Jardine Matheson remained listed among Hong Kong's eighteen largest ratepayers. All the remainder were Chinese.[65] By the 1880s, Chinese became the largest real estate owners in the colony; Chinese investments in land were 7.5 times higher than those of Europeans.[66] Hong Kong was regarded as a land of commercial opportunity, even though the Chinese had no political power, very few legal rights, and suffered humiliating discrimination.

Since the 1840s, schools had been established in Hong Kong by the Morrison Education Society, the Roman Catholic church, the Church of England and the London Missionary Society. The main objective of these mission schools was to train candidates for the ministry, therefore they had little success. The middle-class Chinese who could afford to pay for schooling had little interest in western education.[67]

In 1858, Reverend (later Dr.) James Legge, Senior Missionary of the London Missionary Society, translator of the Chinese classics and the first professor of Chinese at the University of Oxford, went to Canton to visit the examination hall, where scholars were incarcerated in cells for the provincial examination. He was immensely impressed. He believed that there was no other country in the world in which scholastic excellence was as highly revered as in China.[68]

During the 1850s, there was a decline in the number of mission schools, despite the increase in population in Hong Kong. There was not only mistrust between the Chinese and the British citizens in Hong Kong, but a great deal of hostility brought about by the Anglo-Chinese Wars. Dr. Legge believed that the state schools should be secular and provide what was acceptable to the Chinese, in order to attract them to a western education, and in time, to win their cooperation.[69]

It was against the background of hostility and the failure of the mission schools that the Central School was founded in 1861 by Dr. James Legge, a "grand central school" which consolidated all city schools. Dr. Legge realized that the Chinese merchants who brought their families to live in Hong Kong in the 1850s would soon become aware of the commercial advantages of learning English. Frederick (later Dr. Frederick) Stewart was nominated by Dr. Legge as its first headmaster. The school combined liberal views with a knowledge of China and sympathy for Chinese aspirations. In this school, boys of different races worked harmoniously together. The school took pride in the many times when the headmaster received the special messenger from the Imperial court in Peking, announcing the high marks gained by some former Central School scholar in the Imperial Civil Service Examinations.[70]

By 1869 a generation of Hong Kong Chinese who were not only fluent in English, but were experienced in Western methods of business and politics, was emerging.

The name of the Central School was changed to Victoria College in 1881 and to Queen's College in 1894. The school took great pride in the fact that Sun Yat-sen, the father of the Republic of China, was one of its students. He enrolled as student number 2746, on April 15, 1884, under his student name of Sun Tai Tseung.[71] Queen's College came to be regarded as the Harrow or the Eton of the colonies. The school was held in honour the world over. Many of its boys became prominent citizens, and my grandfather was one of them.[72]

Realizing the opportunities for his family in Hong Kong, Great-grandfather enrolled his second son in Queen's College. At the beginning of 1897, the year of Queen Victoria's Diamond Jubilee, a large number of boys were admitted into Queen's College. Grandfather Ting Sien was student number 6707.[73]

Aware of the great heritage of the Chinese culture, and wishing to bridge the gap between East and West, the Central School was first intended to admit only Chinese students and English was not compulsory. However, the school attracted boys from many different races and religions. The diversity was obvious when Chinese boys wearing long gowns and queues, Indian boys wearing turbans, and European boys in jackets and ties were all studying and working together. By the 1880s, the school had boys from twelve different nationalities. No boy was obliged to contravene the social or religious customs of his race — Muslims had leave during Ramadan, Chinese at Ching Ming, Jews at the Feast of Tabernacles, and Christians at Easter.[74] It was not unusual to see Chinese boys translating Aesop's fables, and boys of other nationalities reciting the old Chinese Three Character Classic. In *The Yellow Dragon* (the Queen's College magazine), the words of Confucius were proclaimed monthly: "All within the Four Seas are Brothers." The students came to understand each other's point of view and outlook in life.[75]

From being a student, Grandfather became a pupil teacher. Chinese teachers in those days did not have the benefit of attending a university or a teacher's college. They all apprenticed as pupil teachers on probation. Early every morning these pupil teachers were taught English and School Method, known as Pedagogy, for two hours. For the rest of the day, they taught the younger students in classes of about sixty students each. The curriculum consisted of the three Rs in Chinese and English, and the Scriptures, not as a religion, but for translation purposes only, and only with the parents' permission.

The pupil teachers sometimes found themselves teaching in the assembly hall concurrently with two other classes without screens separating them and without platforms to stand on. It must have been nerve-wracking coping with sixty students under these circumstances. Their evenings were spent studying. Because there was no boarding, and because they received a monthly allowance of only ten to fifteen dollars, the pupil teachers could not afford to take a tram or a sedan-chair. They walked to school every day.

When Grandfather attended Queen's College, he lived above Lai Cheung Loong (a company started by Great-grandfather and his brother Man Yik), on Queen's Road Central. The walk to school was not too far. During the holidays, Grandfather went back to China to his family in Garlieu. During Great-grandfather's visits to Hong Kong, he also stayed at Lai Cheung Loong.

Some pupil teachers went on to become Chinese masters at Queen's College, some later held important posts in China, and some became leading citizens of Hong Kong, as in the case of Grandfather.[76]

In 1898, Grandfather had attained the marriageable age of eighteen. My great-grandparents, according to Chinese custom, contacted the village matchmaker to proceed with the arrangements. The matchmaker would bring together families that were equal in status and wealth. As the saying goes: bamboo door matches bamboo door, wooden door matches wooden door. The horoscopes of the young man and his prospective bride would then be read to make sure they were compatible. Thus Grandfather's marriage to Wong Lan Fong from the village of Toisan was arranged. The ceremony was held in Sunwui at the village of Garlieu. He was still attending Queen's College at the time.

In 1900, on Prize Giving Day, Grandfather was among the recipients for special prizes.[77] In 1901, he left Queen's College to join the workforce. However, he continued to be very much involved with his old school. In 1910, together with two other old boys, he gave generously to the Queen's College St. John Ambulance Brigade. In the same year, he also founded the Lee Hysan and Grant Scholarships.[78]

In 1925, Grandfather agreed to the provision of a further scholarship at Queen's College in memory of his friend Kong Ki Fai.[79]

In 1920, the Amateur Dramatic Association of Queen's College was revived, and several performances of *Romeo and Juliet* were mounted, with the proceeds going to the famine and war distress funds. Grandfather gave medals to the boys who sold the most tickets. He also paid for the school band to play at the Hong Kong Schools Sports Day. Because of his generous contributions, his portrait was hung with that of Bertram Tanner, the headmaster, at Queen's College for the year.[80]

The Queen's College Old Boys' Association was formed in December 1920, to help the school and to provide a link between past and present scholars. Grandfather was one of the pioneers commemorated in a rhyme composed by Mr. Kay, the editor of *The Yellow Dragon,* the college magazine, just before the second reunion dinner the following year. "The mainspring of the OBA since ever it began, deserves the thanks of all Old Boys, so drain your glasses dry, And drink success to Anderson and his 'Anti Mui-Tsai.' Our Treasurer, Barretto, is a good and honest man. Tomorrow morning when you have fat heads blame Lee Hysan..."[81] Every year grandfather supplied the champagne at the Old Boys' Dinners, generally known as "Lee Hysan Joy Juice."[82] He became president of the Association in 1928.

Lee Hysan

According to Chinese custom, a man often has two names in his lifetime. He is given a name at birth by his parents, and he gives himself another one when he goes to work or when he gets married. When Grandfather started his business career, he gave himself the name Hysan. The name Ting Sien was not used again except by his parents and in the ancestral records. The word Hy means hope and San means prudence, meaning he hoped to conduct his life with prudence.

Grandfather was a tall, robust and distinguished-looking man with a classic Chinese face, dark skin, slanted eyes and high cheekbones. Although he was well versed in Western ways of thinking and doing business and was fluent in English, he always dressed in Chinese attire, and always followed Chinese customs. He was intelligent, hard-working, enterprising, and well-liked by his peers. He was also a man of great generosity.

His first job was in a firm in Bonham Strand called Sui Wing Cheung in 1901. The proprietor was Chan Niem Deem. The company imported Chinese medicines and dried foods from China and re-exported them to southeast Asia. Grandfather's talents made him so successful that he was promoted to the position of manager when he was still only twenty-one.

Grandfather's next position was with Seung Duk Fung, a shipping firm owned by the Singapore Chinese Trading Co. Business was conducted between Malaya and Hong Kong, and he made many trips on their vessels. Since he was always looking for business opportunities, he subsequently went into the shipping business for himself, buying Nam Hung Shipping Co., whose cargo ships plied the seas between Hong Kong, Singapore, Malaya and Rangoon. Within a few years, he became very well known and respected by the merchants of Nam Pak Hong, who conducted import and export business between China and southeast Asia.

Grandfather benefited from the entrepreneurial spirit and hard work of the Chinese merchants who had come before him. With the establishment of different associations from the 1850s on, especially the Nam Pak Hong Business Association, the Chinese were able to help each other both in business and in providing security for the Chinese community.[83]

The Chinese in Hong Kong in the early part of the nineteenth century had been completely segregated from the Europeans. In 1864, a policy was established that no Chinese, no sedan chairs and no dogs not on leashes were allowed in the park. No Chinese were allowed into the library or the City Hall museum at the same time as Europeans.[84] The police existed for the security of the Europeans and their businesses only. The Chinese therefore had to form their own associations to protect their homes and businesses by hiring security guards, and the wealthier members donated money to buy fire engines, for fires were frequent occurrences.[85]

When J. Pope Hennessy became governor in 1877, conditions for the Chinese improved, because he realized that the prosperity of Hong Kong depended on the Chinese population.[86] By the 1880s, Chinese merchants had become an important component in the economy of Hong Kong. Not only were they the largest landowners, they also paid 90 percent of the taxes in the colony.[87] They had become major competitors to the European companies.

Land in the central of part of the colony had been reserved for the Europeans until Hennessy became governor. When great-grandfather Leung Yik and great-grand-uncle Man Yik returned from San Francisco with the boys in 1896, Man Yik returned to the village of Hoiping and Leung Yik moved our family to Sunwui. Both brothers then went to Hong Kong to start the company Lai Cheung Loong at 202 Queen's Road Central, and Leung Yik enrolled Grandfather at Queen's College at the beginning of the following year. Both brothers commuted between the villages in south China and Hong Kong. After he went into business for himself, Grandfather took over Lai Cheung Loong by injecting more capital and by buying the property.[88]

Lai Cheung Loong imported blue cotton from Shanghai to sell to the Chinese in Hong Kong. Blue cotton was very popular for everyday clothing for working people at the time. Grandfather used the second floor of the building for his office. Family members who were visiting Hong Kong, such as Great-grandfather or other relatives, would stay in the upstairs quarters on the third floor. Grandfather's younger cousin Lee Sui Yuen also lived there when he was attending Queen's College.[89] The company continued in business until just before the end of the Second World War, and the property remained in the Lee family until 1992.

On weekday mornings, Grandfather usually visited his friends at the various banks, such as the Hong Kong and Shanghai Bank. An old Queen's College friend, Ho Wing Cheun, was the comprador of the Banque de l'Indochine, and another old school friend, Yiu Gueh Yin, was at Citibank, at the corner of Ice House Street. These banks financed Grandfather's various business ventures. Even though he was fluent in English and could have dealt directly with the European managers, he preferred to give the business to his friends.

Later in the morning, Grandfather would drop into Lai Cheung Loong to check his mail, and to look after business at the office. Between 12:30 p.m. and 1 p.m., he would have lunch at the Yue Kee Chinese Club, at 196-198 Wellington Street, where he would discuss business with his friends. Afterwards, he would drop into the stock exchange to keep an eye on his many investments. The brokerage house he usually dealt with was Benjamin & Pots. He would discuss the stock market with brokers such as Sze Yee Man and Go Bo Sung. He invested heavily in many public companies, including China Light and Power Company, China Sugar Refinery, the Dairy Farm Ice and Cold Storage Co., Hong Kong Electric Co., Hong Kong and Shanghai Bank, Hong Kong Tramways Ltd., China Provident Loan and Mortgage Co., and Green Island Cement Co.[90] In the evening, he would often go back to the Yue Kee club for dinner, since this was the place where a lot of business was conducted. He never had dinner at home during the week. He would return home around 10 p.m.[91]

One of Grandfather's investments involved the opium trade, which was legalized in Hong Kong in 1858. Every year the *Government Gazette* called for tenders "for the privilege of preparing and selling opium within the colony pursuant to the 1858 Ordinance."[92] In April 1912, the Yue

Hing Company was incorporated by the Ma family from Sze Yup district in south China, with a capital of $200,000 made up of 2,000 shares of $100 each. The major shareholders were Ma Chi Loong and Ma Jui Chiu. Grandfather also bought shares with a number of others. Yue Hing Co. generally dealt in the opium market. In March 1914, a petition was presented to the court on behalf of one of the shareholders, who was apparently dissatisfied with the way things were being run, asking for the company to be wound up. The Official Receiver was appointed Provisional Liquidator, but the winding-up petition was vigorously contested by other shareholders. The petition was heard by the Chief Justice Sir William Rees Davies. The judgment was eventually given in favour of the company.[93]

In 1913, Grandfather was elected as a director of the board of the Tung Wah hospital. The hospital had been opened in 1872 with the assistance of the government and the wealthy Chinese in the community and overseas. It was solely for destitute and dying Chinese. In 1894, when many Chinese had died of the plague, the Hong Kong government blamed the Tung Wah Hospital Board for mismanagement. Thereafter, doctors trained in western medicine joined the Chinese herbalists on the staff at the hospital. The government took control of the hospital in 1903, providing it with an annual grant of $6,000.[94] A committee of Chinese directors was elected annually to manage the hospital. Each director represented a different trade. The Tung Wah Committee became the focus of Chinese power in the colony, intimately linked with the neighbourhood kaifongs and the merchant guilds. The board looked after not only medical but also the social issues. For the first time, the colonial authorities had a reliable means of canvassing collective Chinese opinion.[95]

In 1923, Grandfather became one of the four members of the consulting committee of the China Light and Power Company, succeeding Dr. J.W. Noble.[96] Fung Wa-chuen, the comprador of Shewan Tomes, had founded the Canton Electric and Fire Extinguishing Company in 1898 with a few of his friends, having gained the concession from the viceroy of Canton province for lighting Canton and the surrounding area. The company ran into financial difficulties and approached Shewan Tomes for help, resulting in the registration in 1900 of China Light and Power Syndicate by Robert G. Shewan with a capital of $200,000 divided into two hundred shares of $1,000 each.[97] In January 1901, the Consulting Committee, which was set up to guide the syndicate, decided that the available capital was insufficient, and that a company should be formed to take over the business. On January 25, 1901, the China Light and Power Company was incorporated. Grandfather was a major shareholder in the company; by 1928 he owned 43,416 shares.[98] The Canton plans proved abortive and in 1909, that operation was sold to a Chinese company.[99] China Light and Power Company remains one of the most important utilities companies today.

Grandfather was keenly interested in the youngsters of Hong Kong. He was responsible for getting the government to grant a large piece of land at the base of Caroline Hill to be used by the public for athletics. He became the first president of the South China Athletic Association.

On June 6, 1925, the foundation stone of St. Paul's Girls' College was laid by Reginald Edward Stubbs, Governor of Hong Kong, with the aim of furthering the education of Chinese girls.[100] Grandfather donated $30,000 towards the building, which was the second-highest donation after the Hong Kong government's $50,000.[101] Thereafter, the assembly hall was named Lee Hysan Hall. After the Second World War, St. Paul's Girls' College became St. Paul's Co-ed College. I was enrolled in the college in grade three and remained in the school for many years. I have many fond memories of the place. The friends I made remain my very good friends today. St. Paul's Co-ed is well known for its discipline and high academic standing. Today, it remains one of the best schools in Hong Kong.

East Point

Although Grandfather had many business interests, his greatest passion was real estate and real estate developments. He had great insight into the development of Hong Kong. He purchased a lot of land and buildings in the central part of the island and in East Point and carried out large-scale developments.

On November 27, 1923, Lee Hysan Estate Company Limited was registered at 202 Queen's Road Central. The telephone number was 2145 and the telegraphic address was "LeeHysan HongKong." The registration was witnessed by C.E.H. Beavis, solicitor of Wilkinson & Grist, which to this day remains the legal representative for our company. The company was established to purchase or lease lands and buildings, and to construct, improve, control and manage roads, tramways, dwelling houses, hotels, clubs, restaurants, places of amusement, gardens and shops in the colony of Hong Kong.[102]

Before making any major investment, Grandfather did extensive research. This was what he did when he bought East Point Hill, also known as Jardine Hill. East Point, in the early nineteenth century, was an isolated promontory and was the most beautiful spot on the almost barren island. It was surrounded by hills except on the harbour side, and from there one had a clear view of the mainland. In front, the deep water at Lei Yue Mun is suitable for deep-draught vessels. Lei Yue Mun is a narrow entrance surrounded by steep hills, protecting the harbour from the prevailing easterly winds.

In 1841, the firm Jardine, Matheson and Co., the premier British mercantile house in the Far East, established itself at East Point. Dr. William Jardine, James Matheson and Hollingworth Magniac were trading partners who had set up a business in the 1820s in Macao with interests extending to India. Dr. Jardine, a Scot from Dumfriesshire, had been a surgeon in the service of the East India Company. Matheson was the second son of Captain Donald Matheson of Shinness in Sutherland, where his family had long been established.[103] Magniac descended from a wealthy old French family called de Magnac. They were Huguenots who had escaped to England from France after the Revocation of the Edict of Nantes in 1685.[104] Around 1805, Hollingworth Magniac went to Macao, where his name was connected with an old established firm, Beale, Read and Co. He was probably also a partner in Charles Magniac and Co. After Charles died in 1824, the company became Magniac and Co. and Hollingworth and his brother Daniel were partners.[105]

The firm Magniac and Co. held a licence under the East India Company and acted as agents for William Jardine and James Matheson in Macao.[106] In the early days of this business connection, Jardine made trading voyages between India and China. Matheson remained in India to attend to the disposal of goods from the Far East. Hollingworth Magniac acted as agent for the sale of goods in Canton and Macao. As time went on, the business increased so that in 1827, when Magniac left China, Jardine and Matheson found it necessary to take up permanent residence in Macao, moving north to Canton in the season, as was done in those early days, and conducting their business there through the "licensed" house of Magniac and Co. Jardine became a partner in 1826, and Matheson in 1828. On June 30, 1832, the firm of Magniac and Co. was dissolved, and on July 1, 1832, Jardine, Matheson and Co. was established.[107]

In order to justify his vision that Hong Kong must become a new Canton, Captain Charles Elliot, the Superintendent, had encouraged the sale of land for development. The first fifty sites were auctioned off on June 14, 1841. The haste lead to ambiguities in regard to the type of ownership. Jardine, Matheson and Co. purchased East Point from the Crown, with the assumption that it was to be either on a long lease or freehold. To their dismay, after the purchase, all leases were fixed for the term of seventy-five years. Alexander Matheson's objections resulted in the subsequent granting of a 999-year lease.[108]

Substantial offices, godowns, stables and dwelling houses were erected at East Point. A slipway was laid down for hauling up and repairing the fleet of schooners and brigs employed by the firm in the coastal trade of the day. The godowns abutted on the seafront and ships came alongside. Magnificent dwelling houses were erected for the partners on a hill 200 feet above sea level, overlooking the harbour. The dwellings were surrounded by an unusually large compound with a fine avenue of trees and had wide verandas, spacious and lofty rooms and passages, and finely dressed stone exteriors. Cannons were mounted on the seafront godowns to guard the compound from pirates.

The offices erected at East Point were used by the firm until 1864, when it moved to the central part of the town. The buildings were thereafter used as junior mess quarters. In 1905 Sir Robert Jardine died and the firm was turned into a private limited liability company.

In about 1923, Grandfather decided to buy East Point Hill. He had observed that Kap Shui Mun entrance was very shallow, therefore all large ships entering Hong Kong harbour had to go through Lei Yue Mun. According to harbour experts, who had come out from England around 1920 to investigate the land surrounding the harbour, the development of West Point had reached its limit, and the future of development lay in the east part of the colony, where Kowloon Bay is situated. Grandfather planned to use material from East Point to create housing on reclaimed land off North Point. North Point is in the middle of Kowloon Bay on the Hong Kong side. Given the rate of development in Wanchai and the rise in property values, as well as the growing population, North Point had an immense future. The foreshore at North Point could be reclaimed, using material from East Point Hill, which was the nearest available site from which easily worked material could be obtained. The completion of this scheme would make available for immediate development a large area of ideally located building sites, which, Grandfather calculated, could be developed at a profit. The area to be reclaimed was 636,040 square feet, at a cost of $1,040,369.[109]

According to Grandfather's calculations, 467 Chinese apartment houses could be built on the reclaimed land. Because of mass production, the houses could be built more cheaply than those built by contractors in Hong Kong. Iron and cement would be imported under contract. Timber and brick would be bought in large quantities. The cost of each house was estimated at $11,400. Each flat could be let at $20 a month, which would yield an 8 percent rate of interest from the investment when fully let. Since there were no reinforced concrete houses with sanitary plumbing available in the colony at such a low rental, the houses would be very attractive to middle-class tenants.[110]

On January 17, 1924, Lee Hysan Estate Company Limited bought East Point Hill from John William Buchanan Jardine, Esq., of Comlongon Castle, Dumfriesshire, Scotland, for the sum of $3,850,960.35.[111] At the same time, another prominent Hong Kong businessman bought land in West Point. He made it known that he thought Grandfather had made an error of judgment in buying East Point Hill. However, history has proved him wrong.[112] The purchase of East Point Hill must have been one of the high points of Grandfather's career. Purchasing the original homestead, offices, and godowns of the taipans of Jardine Matheson represented what the Chinese had been able to accomplish in Hong Kong despite great adversity.

When Grandfather purchased East Point Hill, he had an agreement with the Hong Kong government to use the soil of the hill for the reclamation off North Point. This would give him the right to the development of the reclaimed land. However, after the purchase, the government reneged on the agreement, and decided instead to use the soil from Morrison Hill, which was government-owned; the reclaimed land would therefore remain in government hands. Subsequently, it was found that Morrison Hill consisted mainly of rocks, which had to be blasted at tremendous cost in time and money.[113]

In the meantime, in order for East Point Hill to produce an income, Grandfather turned it into a garden and amusement park, known as Lee Gardens. The taipans' mansions became restaurants. Lee Gardens became a year-round pleasure ground for the Chinese and was financially very successful.

In 1925, the majority of the shares of the Lee Hysan Estate Company were assigned to members of the family. Grandfather had 1000 shares. Each son was given 500 shares. Grandmother Wong Lan Fong had 250 shares. Grandfather's three concubines, Lee Cheung Hee, Lee So Han, and Lee Ng Yuet were given 100 shares each. Each daughter received 70 shares. Smaller share allotments of 5, 4, 2 and 1, were given to various relatives.[114]

The formation of an estate company was a very wise decision on the part of Grandfather, who is, to this day, respected by his descendants as a man of vision. Since the family members are all shareholders, they must work together for the common good. Subsequent amendments were made to the articles of association by the descendants to make the share distribution more equitable.

Grandfather also bought the land at the base of East Point Hill. The slums were cleared, well-constructed houses were built and wide streets laid out. In 1926, the stately Lee Theatre was erected on Percival Street, with a revolving stage for Chinese opera, which was the most popular form of entertainment for the Chinese in those days. It was the most progressive theatre of its time in Hong Kong. The centre of the theatre had a spectacular high dome decorated with dragons and lights. A competition was held, and the winner, Li Ching Yat, a Chinese scholar and teacher at Queen's College, was selected to write the Chinese characters which adorned the theatre.[115]

The Lee Theatre site has since been redeveloped as the Lee Theatre Plaza, completed in 1994. A replica of the original dome of the theatre has been placed on top of the new building. The Chinese couplets written by Li Ching Yat, which were put away for safekeeping during construction, now adorn the foyer of the new Lee Theatre Plaza.

Grandfather continued to amass real estate holdings. Some were purchased in his own name, some in Grandmother's name, and some in the name of Lee Hysan Estate Company Limited. In addition to East Point Hill, our family home on Kennedy Road, and his office on Queen's Road Central, he owned many buildings and apartments on Percival Street, Lee Tung Street, Wanchai Road, Tai Wo Street, Stone Nullah Lane, Queen's Road East, Praya East, Second Street, and Spring Garden Lane.[116]

After the Second World War, East Point Hill was gradually levelled for development, and the soil was used for reclamation in different parts of Hong Kong. The present boundaries of the original East Point Hill are Jardine's Crescent on the east, Leighton Road on the south, Lee Garden Road on the west and Hennessy Road on the north. Today the area and beyond is largely owned by Hysan Development Company and comprises Caroline Centre, AIA Plaza, 111 Leighton Road, Hennessy Centre, Sunning Plaza, Sunning Court, Leighton Centre, One Hysan Avenue, Lee Theatre Plaza and Lee Gardens Redevelopment.

Family

Grandmother Lee Wong Lan Fong was the matriarch of the family. She was born on the third day of the sixth moon in 1880, which made her three months older than Grandfather. She was a fairly tall, big-boned woman with a round face. In the first few years of their marriage, she was not able to have children. She gave birth to a daughter who died in infancy.

Grandmother had bound feet because she came from a fairly well-off family. Foot binding was a mark of social status to show that girls from such families did not need to work. The feet of baby girls were bound to stop their growth; as they grew the bones were broken in order to maintain the ideal size of four inches from heel to toe. Girls had to endure excruciating pain because of their family status. Chinese men were supposed to find small feet erotic. When these girls grew up, they walked like a branch of a willow tree wavering in the wind, which was supposed to be very feminine. I will always remember how difficult it was for Grandmother to walk, even though the bindings on her feet were later removed. Her feet were so misshapen that they were quite grotesque.

Even though Grandfather had a Western education, he was very traditional in many ways. Unlike many men at that time, who abandoned their wives by arranged marriage, he accepted his wife's position as supreme in the family. He respected Grandmother and put her on a pedestal. As we can see in the old photographs, unless it was a group photograph, he would be photographed only with her or with her and the children. When he purchased properties, a number of them were put in Grandmother's name. Grandfather was an opium smoker, but when he found out that Grandmother was allergic to the smoke, he stopped smoking at home.

Although Grandmother was illiterate, Grandfather hired a teacher to teach her to sign her name in English. Apparently the lessons went on for a very long time. She called English writing the drawing of chicken intestines. After the Second World War, I met the English teacher, an East Indian lady who used to come to pay her respects to Grandmother at Chinese New Year.

I knew Grandmother as a very frugal person. She never smiled and never raised her voice. After the Second World War, whenever we went to see her, and that was very often because we all lived on the same family property, she was always sitting in the same chair, looking very unhappy.

Grandfather took a concubine, Cheung Hee, from the city of Canton, who became Yee Tai or Second Lady. When a concubine is brought into the home, she has to "kowtow" to the wife and serve her tea. From then on, she is part of the family. Second Lady was a petite woman, very slender, with a narrow face. She came from a humble family, as all concubines did, therefore she did not have bound feet. Poor girls had to work, so they needed to be steady on their feet. Being a city girl, she was more knowledgeable and worldly than Grandmother.[117] Aside from that, no one seemed to know anything about her family. Concubines were not allowed to visit their own families, or have their families visit them. They were chattels that belonged to the household. Second Lady accepted that custom, never rebelled, never talked back. As a result, no one, including her own daughters, knew anything about her background except the fact that she was about seven years younger than Grandfather. She was a real survivor in that system. In time, she gained respect from everyone in the family.

When Second Lady became pregnant, there was great excitement in the family. Grandfather arranged to have a European midwife for the delivery.[118] It was Grandfather's first child after seven years of marriage. I guess he felt he couldn't take a chance on this baby, and believed that a European midwife would be more knowledgeable and more hygienic than a Chinese midwife. On March 7, 1905, Second Lady gave birth to the first son, Ming Chak, my father.

Dad's birth was looked upon as a good omen for the family. Not only because he was a son, but after his birth, Grandfather's business prospered. He was placed in the care of Grandmother, in the hope that this would bring her luck in having sons of her own. She did subsequently give birth to two sons and two daughters: third son Wing Gun (later known as Hsiao Wo), fourth son Jung Sen (J.S.), second daughter Shun Wah and fifth daughter Shun Yin. A special relationship developed between Dad and Grandmother that lasted a lifetime. She became mother to him. She respected his judgment, which later benefited the entire family.

Second Lady later gave birth to second son Ming Hop and subsequently three daughters: third daughter Shun Ying, fourth daughter Shun Kum and sixth daughter Shun Ho.

Grandmother was the head of the household, and regarded as the mother to all the children. The concubines not only had to serve the husband but the wife as well. The children of concubines call the wife "mother," and their own mother "aunt." All the children were legitimate and equal under Chinese law. Concubines were sometimes allowed to leave, if permission had been granted by the master of the household, or by the wife after the death of the master, but they could never take their own children with them. The children were part of the family.

Second Lady also respected Grandmother and never argued with her. She would often walk away from a potential argument by saying that Heaven and Earth knew the truth.[119] Nevertheless, I remember Second Lady as a high-spirited person. She used to tell us children that if we were not good, she would personally take the skin off our backs. But we knew she would never dream of doing anything like that. She was a good mother and grandmother. Throughout her life, she did her best, not only for her children, but also for her grandchildren who needed help.[120]

Even though she had servants, for Chinese New Year, Second Lady liked to make steamed cakes herself in the old-fashioned way in the garden. I used to watch her squatting in front of an old Chinese stove, blowing on the straws that she fed into the fire. You would think she was still living in a village in China. The steamed cakes would be distributed to her children's family. Every year, for Dad's birthday by the Chinese calendar (Lunar calendar) she would make him a village dish—duck cooked with taro, because she said that had been his favourite dish as a child.

I used to love watching her smoking her water pipe. She would stuff tobacco into an opening in her beautiful cloisonné pipe, which was filled with water. Then she lit a long paper stick, and held it to the tobacco to burn it. When she inhaled, the smoke would be drawn through the water. The water would make a bubbling sound, then puffs of smoke would come out of her mouth. The gurgling sound used to make us children laugh. It was one of our favourite entertainments.

Second Lady prayed to Buddha every day in her prayer room. She would put on her prayer robe, which was made out of a coarse dark brown cotton, before she knelt in front of the altar. In addition to the statue of Buddha, there were statues of Chinese deities on the altar. She would burn incense, and bring fruit and whatever she was having for dinner that day, which had to be offered to Buddha first. She prayed while counting her prayer beads. The prayer seemed very repetitious, almost like a chant. As children, we used to run in and out of her prayer room while she was praying and make faces behind her back. She would normally get after us for doing something like that, but she never noticed when she was in prayer.

In her later years, Second Lady gave all the jewellery that Grandfather had bought her to her daughters. She said she had no need of it. If there was a special family function at which she needed to wear jewellery, she would borrow it back.[121] After her seventieth birthday, she divided all her savings (savings from her housekeeping money, because she never had money of her own) and gave everything to her descendants. I remember receiving $700 from her many years ago. She had her burial clothes made, and prepared herself for an end that did not come until many years later.

After Grandmother died in 1956, Second Lady came to be regarded as the matriarch of the entire family. When she was in her eighties, she made her first trip out of China by flying to San Francisco. Not only did she not know a word of English, but she had never been on a plane before. She said she had no fear because Buddha would look after her. She lived a contented and healthy life. After having smoked a water pipe all her life, she finally died of lung cancer at the age of almost 100.

When Grandfather began to attain wealth and status in Hong Kong, he heard rumours that his parents, who lived in Garlieu, were in danger of being kidnapped by bandits. Grandfather moved them to Sunwui city and built them a house at 28 Juk Lum Lai.[122]

The house was built in the typical design for the wealthy of its day. Above the wooden gate that opens into a square courtyard were beautiful wall paintings. On the right-hand wall of the courtyard, just inside the gate, was an altar for burning incense to Chinese deities (this was destroyed during the Cultural Revolution). Large painted clay goldfish were placed on top of the drainpipes surrounding the courtyard. Rainwater would drain into the tails of the goldfish and pour out of their mouths and down into the pipes.

Opposite the door of the sitting room, ancestral tablets were displayed over a hand-carved altar table inlaid with marble. On top of the altar, incense burners, dishes of fruit and, on special occasions, food, were placed as offerings to the ancestors. Antique armchairs stood on both sides of the sitting room. One of these chairs was supposed to be very special. On one of their trips back to Sunwui in the early 1980s, Dad told cousin Hon Chiu (the eldest son of second uncle Ming Hop, and Grandfather's eldest grandson) to sit in that particular chair, because anyone who did so would be able to manage the family business well. Dad had obviously done so himself on a previous occasion.[123] Cousin Hon Chiu has also proved the belief to be true.

To the right and left off the sitting room were bedrooms, and at the back were the kitchen and the storehouse. There was a well in the back courtyard. The bathroom and washing facilities were housed in a separate structure. At one side of the house was a spacious walled garden. A flight of stairs from the courtyard led to the roof, which commanded a clear view of the entire neighbourhood.

This ancestral home still stands in its original form, despite the damage done by the Communist Revolution and the subsequent Cultural Revolution. Our relatives in Sunwui take care of the house, but it is no longer used as a dwelling.

The family made frequent visits to this house during my great-grandparents' lifetime. When Dad was a little boy, Great-grandfather used to tell him stories of his experiences in San Francisco and of the terrible journey across the Pacific. Since food has always been of great importance to the Chinese, Great-grandfather used to stuff chicken legs into Dad's mouth, even when he was asleep, to show his affection.[124]

When Great-grandfather died, according to the village custom, his body had to travel by boat along the river from the village of Garlieu. Dad sat in the front of the boat, and Grandfather at the back with the coffin in the middle.[125] The burial place that was picked was almost inaccessible. It took a day of travelling to get there, and then the family had to climb a hill to get to the top.[126] The plot was considered to be very lucky for the descendants.

Due to the repeated outbreak of the plague and the insanitary conditions in Hong Kong, Grandfather thought it best to move the family to Macao in 1910. The family lived in a house on Holland Yuen Street until 1918.[127] (The street got its name because it was the site of a Dutch prisoner-of-war camp in the 1620s, during the conflict between Portugal and Holland.) Because education for his children was very important to Grandfather, a private teacher, Chan Tse Po, was hired to teach them during their absence from Hong Kong.[128]

When the family moved back to Hong Kong, my father, Ming Chak, was enrolled at Queen's College. At the age of fifteen, he was sent to school in England with his younger brother, Wing Gun (later known as Hsiao Wo). They lived in the home of their English guardian, Mr. Churchill, and were taught by tutors. Later they were joined by their two sisters, Shun Wah and Shun Ying, who were sent to a boarding school. Shun Wah was the daughter of Grandmother, and Shun Ying was the daughter of Second Lady. The dates of birth of the two girls were less than nine months apart. This caused technical difficulties when the sisters needed travel documents to go to England. Since it would have been too complicated to try to explain the Chinese system of concubinage, the birthdays were altered instead.

The children adopted English names: Ming Chak became Richard Charles, and Wing Gun was called Harold; their sisters were known as Doris and Ansie. Dad (known as Dick to his friends), subsequently went to Oxford to study engineering and uncle Harold went into law.

Around the time of the First World War, Grandfather bought a large piece of land on the side of a hill above Wanchai to build a home for himself and his family. Because of the war and labour problems, building was delayed, but in 1920, the firm Palmer and Turner designed and built the Big House (Dai Uk). The address was 32 Kennedy Road (later changed to 74), and the phone number was 2455. This was another high point of Grandfather's career, to have a house with such a commanding view. In the nineteenth century, this would not have been possible for a Chinese family.

The property on Kennedy Road was surrounded by a high stone wall, which contained gardens of flowers and fruit trees, vegetable gardens, chicken coops, pagodas, fountains, goldfish ponds, and artificial hills and caves. Later on, tennis courts were added. From the garden, there was a breathtaking view of Hong Kong. A team of gardeners looked after the grounds, and there was always an abundance of beautiful flowers.

At the front gate of the Big House was the guard house where tall Sikhs with shotguns stood guard around the clock. People in Hong Kong believed that the most fierce-looking guards were tall Sikhs in turbans. That was because they had beards, whereas the Chinese are a rather hairless race. Our guards and their families lived in apartments on one side of the garden, by the fountain of the Goddess of Mercy. These guards worked for our family for generations.

After the Second World War, the guard I knew the best was Nam Singh. In Chinese, it meant South Star. Little did I know at the time that all Sikhs have Singh as their last name. These guards all spoke perfect Cantonese. They never seemed to have much to do except look fierce with their shotguns. Nam Singh was usually on duty during the day, and he always chatted with us when we played in the garden. Once, he showed me the snakes he caught which he kept under a flower pot. He told me he was going to cook them for dinner when he went off duty.

I have wonderful memories of our home on Kennedy Road; I was born there and our family lived there for many years. Many of the cousins lived on the same property, so my brother, sister and I were never short of playmates. The grounds were so big it was like an adventure every day. Throughout the year, we could pick fresh fruits from the trees. We had just about every type of fruit tree that grows in south China in our gardens. When we wanted to eat a papaya, some of us would hold out an old raincoat, and one of us would rock the tree where there was a ripe papaya. If the rocking did not work, we would poke it with a stick until it fell into the raincoat. It would split open, and we would eat it without ceremony. However, we always made sure we washed ourselves in the garden before going home so that we would not get into trouble for being a mess.

In the spring, we used handkerchiefs to catch tadpoles in the goldfish ponds, and we chased baby chicks around the chicken coop. In the summer, we would catch heat bugs and dragonflies. There is a type of tree bug the size of a large dragonfly that has a long snout. We would catch them and tie threads on their snouts and fly them like kites. Throughout the year, we climbed trees and swung on their branches, picked fruits and pretended to be Tarzan or Jane. We rode our bicycles all over the gardens. We played on the swings and slides, and walked on stilts. At the end of each day, at bath time, when the servants were sent looking for us, my favourite place to hide was one of the artificial caves.

The Big House was one of the grandest homes ever built in Hong Kong. The front hall alone could easily accommodate a reception for 500 guests. The ground floor was reserved for entertaining, with enormous dining and sitting rooms and a large kitchen. At the back were the servants' quarters. The second and third floors were living quarters with sitting rooms and large balconies for the family. There were many guest rooms in the house for visitors from our village who needed a place to stay.

On the weekends, grandfather usually had his meals at home. Family members and relatives from the village would be welcome at the dinner table. Women and small children would dine on the third floor. The men and the older boys dined on the second floor. Meals were served from the same kitchen and at the same time.[129]

The house was filled with objets d'art from all over the world. Most of these were looted during the Second World War when the house was left empty. However, a magnificent ivory tusk from a mammoth escaped the looters. It reclined on a beautifully carved blackwood base. It had originally belonged to a Frenchman in Shameen, Canton, who bought the raw tusk and commissioned an artist, Leung Wai, to carve on it the story of the Three Kingdoms. It took the artist years to finish the detailed figures and scenes. The Frenchman fell on hard times and had to use the tusk as collateral at the Banque de l'Indochine. Through the bank's comprador, Ho Wing Cheun, Grandfather redeemed it in 1926. As a small child playing in the main hall in the Big House, I used to love riding on this tusk as if it were a horse. Thinking back, I am amazed that no adult ever stopped me. This magnificent tusk has now been donated to the museum at the Chinese University of Hong Kong.

There were opium beds in the downstairs hallway that were beautifully carved in Chinese blackwood. We, the grandchildren, used to climb up to one of them and tell stories. Many small children could fit on one bed.

The main hall of the Big House was very bright, but the side entrance to the stairs to the second and third floors was always very dark because of the wood paneling. After the Second World War, when we were playing, the older children would often run up those stairs saying that we were being chased by a ghost. I was the youngest in that group, and therefore the slowest to get to the top of the stairs.

After the construction of The Big House, Grandfather built Lee Building (Lee Hong) on the same property further up the hill. It was a grand-looking building with a wide open central staircase, consisting of six spacious and bright apartments that were rented to Japanese and European tenants. I wonder whether Grandfather knew that one day the building would be lived in by his descendants. Although the two buildings were connected by numerous gardens, we rarely saw the tenants.[130] After the Second World War, when our families were becoming larger, Lee Building was used by the family as well. Even after we moved into the Lee Building, Mom and Dad continued to entertain their business friends in the Big House, and all the children continued to play in the same gardens.

By the time the Big House was built, Grandfather already had two concubines. According to old Chinese custom, a man showed his wealth by the number of concubines and children he had.

Third Lady, So Han, had been a singer in one of the teahouses Grandfather frequented. He took a liking to her and brought her home as a concubine.[131] She had a high voice, a round face and a very pleasant smile.

Grandfather, Grandmother, the concubines and all the children lived together in the Big House. Grandmother lived on the second floor with her children. Second Lady and Third Lady lived on the third floor with their children. Third Lady often quarrelled with Grandmother, and eventually she left the Big House to live with her son Wing Kit (fifth son) in Causeway Bay in her own apartment on Percival Street.[132] She never moved back to Kennedy Road, not even after Grandfather died. I didn't see her more than once or twice a year, so I knew her the least.

After the family moved into the Big House in the early 1920s, Grandfather took another concubine, Ng Yuet, Fourth Lady. Grandfather had met her while he was being entertained with a few wealthy friends in a house of pleasure. The madam told the gentlemen one day that she had a new girl who was a virgin. This girl was available to the highest bidder. Grandfather gave the highest bid. He was obviously infatuated with her to have brought her home as a concubine.[133]

Fourth Lady never lived in the Big House. She had her own living quarters close to Grandfather's office. Grandfather would often visit her during the day. Sometimes fourth son, J.S., would go with Grandfather. He remembers that she was very kind to him whenever he did not feel well. J.S., who remained in Hong Kong to study Chinese, was the only son who used to make the rounds with his father, visiting the different friends and business associates.[134]

Fourth Lady was fairly tall and very slim. After Mom and Dad were married, she and Mom became friends. She often poured her heart out to Mom, who was only ten years her junior, because Mom was the only family member who was not in competition with her. She told Mom that it was her own choice, and not her family's, to work in the house of pleasure where she met Grandfather. It was the only chance for a poor girl to meet and become the concubine of a rich man. She feared poverty, and could not imagine being married to a poor man. I went to visit her a few years ago, when she was over ninety, but still well enough to live at home. During our conversation, I mentioned that her granddaughter was doing very well in Toronto and was very happy, and she asked, "But is she making a lot of money?" I realized then how important money was to her all her life. It was her only security.

When Grandfather died, Fourth Lady was still very young and had the option to leave and marry again. She told Mom that she chose not to because she wanted to be with her children, and her children were part of the family. She turned to Buddhism, became a vegetarian, and befriended many nuns. After the Second World War, she moved into the Lee building with her children, sixth son Jung Kong, seventh son Wing Tat, seventh daughter Shun Yee, and eighth daughter Shun Ngor.

Lee Shiu Yuen, son of Mun Yik, a younger cousin of Grandfather, has fond memories of Grandfather. When he was a student at Queen's College, he would go to the Big House to visit Grandparents on Sundays, and stay for the noonday meal, together with a number of other relatives. He said Grandfather was a stern man who expected good behaviour and diligence from the younger generation, and they were all in awe of him.

After the Lee Theatre was built, Grandfather used to like taking Shiu Yuen to the Lee Theatre after Sunday dinner. He talked to him about the future of theatres, saying that the day would come when people would be able to watch movies lying down. Shiu Yuen took that to mean watching television at home.

Shiu Yuen remembers Grandfather as a man who believed in hard work and education. He could be quite intimidating to those who were not up to his standard. He was, however, very generous to everyone in his family. He not only took care of his parents, but also his siblings, their spouses and children, and all the needy relatives from the village. He provided education for his nephews and nieces and jobs for his brothers. His brothers' wives were all given housekeeping allowances during their lifetime. This continued even after his death by Grandmother to honour his wishes.

Grandfather provided rent-free accommodations to all his relatives. In 1959, when I first came to Canada, I met an old lady who had emigrated from south China with her family to Ottawa many years ago. When I was introduced to her, she told me that one of her relatives was married to one of my Lee relatives in Hong Kong, and that they had been provided with a free apartment by my grandfather. People remember such unusual generosity.

The family lived in the Big House until the Japanese occupation in 1941, when we all went into China. When we returned after the War, Grandmother and her children lived in the Big House, and the rest of us lived in the Lee Building, with the exception of Third Lady and her son Wing Kit, who remained living in the Causeway Bay area.

In the 1950s, Lee Gardens was levelled and developed, roads were built and high-rises went up. My own family moved into the penthouse in Embassy Court on Hysan Avenue, which was the first high-rise we owned. Subsequently, with the completion of Caroline Mansion and Tower Court, more family members moved into the area.

The property on Kennedy Road where the Big House and the Lee Building stood has been redeveloped by Hysan Development Co. into a large residential complex in 1985 known as Bamboo Grove. A number of the Lee family members live there today.

Dick & Esther

When Grandfather sent two sons and two daughters to study in England in the 1920s he told them that they should not even think of marrying a non-Chinese while they were abroad, otherwise they would be automatically disinherited.

In 1923, Dad entered Pembroke College, Oxford, to study civil engineering. Chinese students studying in England in those days were very well treated, and were not subject to the discrimination they encountered at home. They were able to make friends and contacts that were invaluable to them throughout their lives. Many became prominent citizens in their own countries.

In 1925-26, Dad was the President of the Chinese Students' Union of Europe. One of his contemporaries at Oxford was C.C. Chin, who read economics at University College, and was later made chief economic advisor in Peking. Dad always preserved the menus of the Union dinners at which he and his fellow students sketched their plans for a brave new world in China.[135]

In 1927, after graduating from Oxford, Dad returned to Hong Kong for a visit. He was surprised by the anti-Chinese discrimination which he had forgotten about while he was in England. He had become used to riding in England, and wanted to continue to ride at the Jockey Club in Hong Kong. He was refused because he was Chinese. Grandfather immediately made plans to form a Chinese jockey club. That was considered a threat to the Jockey Club, which depended financially on the Chinese population gambling at the races. A decision was made to allow Dad to ride at the club.[136]

Dad also wanted to join the Hong Kong Club, and was turned down because he was Chinese. This was an insult he never forgot. Years later, he was asked to join the club, but he refused the offer.

As Dad was the eldest son, his parents were naturally very anxious that he should find a bride before he returned to England to do his practical training. Many young ladies were introduced to him, but he was not interested. Young ladies were also brought to see Grandmother. I heard that Grandmother said one girl was not suitable because her legs were too hairy!

One day in the summer of 1927, Esther Wong and her sister Sarah were out playing tennis in Kowloon. On their way home, they passed by a beach and saw a large group of young people having a picnic. They noticed that their older sister Julia and a few friends were in the group, so they decided to go down to the beach to say hello.

Dad was one of the young men in the crowd. He approached Julia for an introduction to one of the young girls who had come down to the beach. Julia said, "Don't bother with her. She's just my younger sister." Dad insisted and was introduced to Esther. They spent the afternoon together and Dad drove her home.

For the next few months, with the approval of their families, Dad saw a lot of Esther. Dad enjoyed entertaining and often had a group of friends with them. He often invited her and other friends to the tea dances at the Hong Kong Hotel, even though he did not like to dance.

Esther, my mother, came from a modern Chinese family. Her father, Joseph M. Wong, was fluent in English and was well respected in Hong Kong society. He was the first president of the Chinese Association of Chartered Accountants, formed in 1913. In 1923, he was made Justice of the Peace by the Hong Kong government. On June 12, 1941, he was given the Badge of Honour by King George VI of England. In the first year of the Republic of China, he was awarded the fourth-level "Gar Wo" badge by Dr. Sun Yat-sen. Mom's mother was Jeannie Maxwell, an Eurasian who was famous for her beauty. In those days in Hong Kong, Eurasians either became very European or very Chinese. Jeannie Maxwell Wong was very Chinese. She was an authority on Chinese customs.

The Wongs were Christians. Mom's paternal grandparents were by all accounts wonderful people. I remember hearing a lot of stories of the fun Mom and her siblings and cousins used to have at their orchards and country house in the present-day Prince Edward Road area in Kowloon.

Great-grandfather Wong Chuen was a court intepreter in the service of the Hong Kong government.[137] Great-grandmother Wong was the favourite of the grandchildren, who talked about her all the time. She was a thoroughly modern woman who spoke English and German besides Chinese. She encouraged mixed marriages among her children because she wanted the best-looking grandchildren. She enjoyed singing and being sung to by her grandchildren. Each grandchild who sang a hymn to her would receive a coin. She spent a lot of time doing charitable work, which continued even after she had a stroke and was confined to a wheelchair.[138] In her later years, Great-grandmother Wong got her husband a young concubine when she was tired of repeated pregnancies. There were a total of sixteen children, and my grandfather Wong was the second son.

Mom came from a family of eleven children, all from the same mother. She was educated at the Diocesan Girls School and was fluent in English. The Wong girls (Mom's sisters and her cousins) were reputed to be the prettiest girls in Hong Kong society in the 1920s. Mom was only seventeen when she met Dad and was still in high school, so she was not yet part of the social scene.

Mom's father was one of the first people in Hong Kong to own a motor car. Mom and her sisters used to drive all over Hong Kong. Mom liked to tell stories about being stopped by the police for speeding. She enjoyed tennis and also took flying lessons, although she never obtained a pilot's licence.

One day, Dad confessed to Mom that he was the son of a concubine and not of Grandmother. He wondered if it would matter to their relationship. Because of his special relationship with Grandmother, everyone had taken it for granted that he was her own son. Mom said no, it didn't matter. After all, that was the way traditional Chinese families were in those days, even though her own parents were very different.[139]

It was a whirlwind courtship. Both Grandfather and Grandmother Lee were delighted by Dad's choice. Grandfather was anxious for his eldest son to get married before he went back to England. He was so anxious, in fact, that when Dad proposed, Grandfather and Grandmother Lee also went to visit Grandfather and Grandmother Wong to ensure their approval.

There was just one problem. Mom was only seventeen and was still in high school. She wanted to graduate before getting married. She also felt that she was too young to leave home where she had so much freedom, because she knew she would be marrying into a very traditional family. Grandfather Lee had to use all his persuasive techniques to win her over. He promised her an education at Oxford University when she and Dad returned to England after the wedding. He suggested that she study Portuguese, so that later on she could help him in his business. He also promised her parents a trip with him to Europe the following year to visit them. In fact, he promised Mom everything that she would ever want so that she would not mind leaving home. No wonder she always had such high regard for him!

The wedding date was set for February 28, 1928. The *Hong Kong Daily Press* reported that the Very Rev. A. Swann, Dean of Hong Kong, officiated for the first time at a Chinese wedding ceremony when Mr. Richard Charles Lee and Miss Esther Yewpick Wong were married at St. John's Cathedral.

The cathedral was thronged with European and Chinese friends. Long before the bridal party arrived, every seat was occupied, and many people had to wait outside the building. The bride and groom arrived together in a new Fiat motor car, attended by bridesmaids and flower girls. The bride looked charming in a gown of silver lace trimmed with pearls, carrying a bouquet of white roses. The maid of honour, the bridesmaids and the flower girls all wore crepe satin of various colours trimmed with rosettes, and carried bouquets of pink roses. The parents of the bride and groom were in full Chinese court dress.

The reception was held at the Lee Gardens and attended by more than 2,000 European and Chinese guests. A huge matshed (a temporary structure built with bamboo and straw mats) was erected for the purpose. The catering was done by the Hong Kong Hotel. A dais was erected in the matshed for the six-tier wedding cake.

The health and happiness of the bride and groom was toasted by Sir Robert Ho Tung, who was not only a good friend of Grandfather Lee and Grandfather Wong, but was also a good friend of Great-grandfather Wong Chuen. In his elaborate speech, he referred to the speech Dad had made as the President of the Central Union of Chinese Students in Great Britain and Ireland regarding his sentiments towards China. Little did Sir Robert know that it would turn out to be a prophecy that Dad did later give his services and knowledge freely for the good of China.

The health of the parents of the bridal pair was proposed by Dr. Robert H. Kotewall, a member of the Legislative Council. The task should have fallen on Sir Shou-san Chow, but unfortunately Sir Shou-san was indisposed and shortly after arriving at the reception had to return home.

Dr. Kotewall was only too glad to fill in, as he had known both families for a long time. He spoke of Grandfather Lee and Grandfather Wong as having given invaluable services to Hong Kong, and said that the mothers of both the bride and bridegroom had been model wives. He said he felt sure that Dad and Mom would go hand-in-hand through life full of happiness.

Dad thanked Sir Robert Ho Tung and Dr. Kotewall. He said he was overwhelmed with gratitude by the kind words that were said, and that he and Mom were overjoyed at seeing so many good friends attending their reception. This would forever remain a red-letter day for them. Grandfather Lee and Grandfather Wong also expressed gratitude for the kindness shown them by the presence of so many friends.

For their honeymoon, Mom and Dad went by boat to Switzerland via the Suez Canal. They had a wonderful time in Egypt and did all the touristy things, such as riding on camels, and taking pictures in front of the Sphinx. When they arrived in Switzerland, they had a family reunion with Dad's brother Harold and sisters Doris and Ansie. After their honeymoon, they went to England.

Trial

In the latter part of the 1920s, Grandfather got into legal difficulties over one of his businesses. With the graduation of his eldest son from Oxford and his subsequent return to visit the family in 1927, the last thing on Grandfather's mind was the winding up of the Yue Sing firm.

For three years before the end of June 1927, the Yue Sing firm held the opium monopoly from the government of Macao, with the right to prepare and retain opium in that colony, and to import and export. Throughout the nineteenth and the first part of the twentieth centuries, not only was the Hong Kong government involved in the opium trade, but also the Portuguese government in Macao. These governments sold licences to companies which dealt in opium; there were also government-licensed opium dens. The governments were in competition with the smugglers, who sold their stock at a much lower price.

In 1924, the Yue Sing firm was formed under the sole proprietorship of Lee Yue Sing, Grandfather's older brother's son. (This firm was not in any way connected to the Yue Hing Company owned by the Ma family which had wound up many years before.) The capital for Yue Sing was $3 million, of which a third was owned by the Lee family. Grandfather was the general manager. In the three years Yue Sing had the opium farm, a quarter of the original capital of $3 million had been returned to the subscribers.

According to the Geneva Convention of 1924-25, and article VI of the Opium Agreement signed on February 11, 1925, the right to an opium farm was not to be leased to any person whatever, and the export of opium whether raw or prepared from any territory was to be prohibited.

In the *Boletin Official* of March 26, 1927, the Portuguese government announced that under the agreement of the Opium Conference held in

Geneva between November 2, 1924, and February 19, 1925, and ratified on March 2, 1926, the monopoly system under which opium had been imported, prepared, sold and distributed would come to an end. As of July 1, 1927, the contract with the Yue Sing firm would terminate, and a government monopoly would be established under the superintendence of the Inspector of Consumption Taxes. Proceedings were initiated by the courts in Macao to get the Yue Sing firm wound up, and the rest of the investment was lost when the farm was taken over by the government of Macao.

The Inspector of Consumption Taxes established an Opium Adminstration and Pedro Jose Lobo was appointed Opium Administrator.

In June 1927, a rumour came to Grandfather's ears that the opium business in Macao had been given to Lu Kin Butt, the manager who had worked under Grandfather in the Yue Sing firm in Macao. In addition to Lu Kin Butt, two others had been appointed; both were smugglers well known to the police. Grandfather subsequently heard that a company by the name of Yau Sing had been formed with an office in the same building as the new Opium Administration in Macao. The company had also opened an office in Hong Kong to invite subscriptions in Hong Kong, Macao and Canton, claiming that it had obtained the contract for the opium monopoly from the Macao government.

The Yau Sing Company had also opened an account with the branch of the Mercantile Bank of India in Hong Kong. Apparently, the information reached Grandfather as a result of a mistake. A messenger from the bank brought him a cheque chopped (endorsed) by the Yau Sing Company. This messenger had been told by the bank comprador to take the cheque to the Macao Opium Farm. On making further enquiries, Grandfather learnt that a few days after the account was opened, the comprador was informed that the Yau Sing Company had obtained by tender the opium monopoly from the government of Macao.

In August 1927, a friend of Grandfather's, Fung Cheung, who had contributed to the capital of the Yue Sing firm, came and asked Grandfather's advice about buying shares in the Yau Sing Company. Grandfather advised Fung to make enquiries about the company. Fung wrote to request information in order to have the answers in writing.

Grandfather was surprised by the information he had received about the actions of the government of Macao. He believed that since the contract with Yue Sing firm had been terminated by the Macao government, then no other firm should legally be given a new contract by the same government. He therefore prepared a petition to the governor of Macao. In the petition, he described his services and benefactions to the colony of Macao, and his fulfillment of all his obligations to that government, and asked for fair and sympathetic treatment where his interests and that of the Yue Sing firm were concerned. He mentioned the public announcement of the future policy of the government of Macao toward the opium trade; the actions taken by the government against the Yue Sing firm and its representative in Macao, which he claimed were unwarranted under Portuguese law; the engagement of Lu Kin Butt by the Opium Administrator without Grandfather's knowledge and Lobo's attempt to shield Lu Kin Butt from criminal proceedings in Hong Kong; the failure of the government of Macao to return, at the due time, the money deposited by the Yue Sing firm with the government; the activities of the Yau Sing Company in soliciting contributions to its capital from members of the public by claiming that it had obtained the opium concession in Macao; and several other matters. The petition ended with a request for an enquiry into the matters.

He sent the petition to the governor, the members of the Legislative Council, sixteen lawyers in Macao, and the Portuguese Consul-General, Senor Albuquerque, in Hong Kong. He also sent it to Cheung Choi, who translated the response to Fung's letter into Chinese.

In spring 1928, Pedro Jose Lobo of Macao sued Grandfather for libel and asked the court for an injunction to prevent him sending a further petition to the governor of Macao.

The alleged libel related to the Macao opium monopoly and stated, among other things, that the Yau Sing Company of Macao had paid $120,000 to acquire control of the monopoly. The plaintiff's case was that this reflected discredit on himself as a government officer connected with opium administration.

The hearing at the Supreme Court was presided over by the Chief Justice, Sir Henry Gollan. The plaintiff was represented by Mr. C.G. Alabaster, K.C., and Mr. G.H. Sheldon. Mr. Eldon Potter, K.C., and Mr. F.C. Jenkin represented the defendant.

Mr. Alabaster asked why Grandfather had had 800 copies of the petition printed and whether he was planning to distribute it worldwide. Grandfather answered that if he had had fewer printed it would still have cost him $400.

The letter from Fung Cheung to Grandfather was read into the record. It stated that "according to the announcement of the Yau Sing Company to the shareholders, the sum of $70,000 was paid through Ah Nok's hands as security and $50,000 was paid to the Treasury through Lobo's hands. The Yau Sing Company has acquired this opium monopoly not by successful tender in open competition, but through the influence of certain persons. It was Ah Nok and the new opium officer Lobo who used their influence to put matters through. The Government of Macao allowed this company to do what they liked. It is normally a monopoly run according to law."

On April 17, 1928, Sir Henry Gollan, Chief Justice of the Supreme Court, gave judgment for Grandfather. In his judgment, his Lordship found that the letter complained of was defamatory to the plaintiff, but its publication in each instance was privileged. He also noted that the plaintiff had failed to prove that in making such publication the defendant was motivated by malice. Grandfather was relieved that the court case was over and was hoping to see justice done by the governor of Macao.[140]

During the period leading up to the trial, Grandfather had received letters threatening his life in order to stop the petition, and even threats that bombs would be thrown at his son's wedding. He had disregarded them, although his friends warned him to be careful. His cousin Lee Shiu Yuen said that Grandfather was not in the least bit worried, and thought that his friends were making a fuss over nothing. If he had taken the threats seriously, he would at least have changed his daily routine. Now two of his sons were married (second son Ming Hop had been married soon after Mom and Dad's wedding in a very traditional Chinese ceremony),[141] and everything had proceeded smoothly. Grandfather was looking forward to a great future.

Heirs

On April 30, 1928, when Grandfather went to the Yue Kee club for tiffin at his usual time, he was assassinated. He was forty-seven. He was survived by his wife, three concubines, seven sons and seven daughters. The youngest child, Shun Ngor, daughter of Fourth Lady, was born after he died. The South China Athletic Association flew its flag at half-mast as a mark of respect for their late president. The Lee Theatre and the Lee Gardens were closed to the public for forty-nine days in mourning for the proprietor. His body lay in the Lee Gardens where many friends called to pay their respects. According to old Chinese custom, the body had to be accompanied by the spouse, concubines and descendants for forty-nine days, during which time prayers are said by monks from different monasteries to ensure smooth transition into the next world.

Grandfather was a well-known figure among Chinese business circles in Nam Pak Hong and was one of the largest landowners in the colony. He was well liked by foreign residents for his charming manners.

The family offered a reward of $10,000 for information leading to the arrest of the assassin. It was reported on May 1 that a suspect had been arrested at the Macao wharf, but the rumour could not be confirmed. Although the police had the slenderest of clues on which to proceed, they did not relax their efforts in the hunt for the murderer. Some of the witnesses had insisted that a man and a boy had been seen leaving the club at the time of the shooting, but they could not give any description of the man. Family members showed great anxiety over the police enquiries, and some of them went to the Central Police station to find out more about the investigation. There were repeated rumours of arrests, but the case was never solved.[142]

Funeral arrangements were announced in the *South China Morning Post* on May 24 for the following day. The procession was to leave the deceased's residence at 74 Kennedy Road (the Big House) at ten o'clock in the morning. It would travel along Queen's Road East to Queen's Road Central, past the bank and the old fire station, past Lai Cheong Loong at 202 Queen's Road Central, to the Pavilion in Kennedy Town where friends would be able to pay their last respects between one and two o'clock. The interment was to be at the Chinese Permanent Cemetery in Aberdeen, Hong Kong.

Dad and Mom, brother Harold, and sisters Doris and Ansie were in England when they received the news of the death of their father. They returned to Hong Kong immediately. Because the only mode of transport was by boat, they could not be present at the funeral.

Grandmother was the sole executrix of Grandfather's will, but since she had no knowledge of business, she depended entirely on Dad. With this sudden turn of events, Dad, at the age of twenty-three, became the head of the family. His immediate concern was to raise large sums for the payment of death duties. Just about all of Grandfather's shares had to be sold. The largest block of shares was that of China Light and Power, bought by Grandfather's friend Sir Elly Kadoorie, who joined the board shortly afterwards.[143] For the first time in his life, Dad had to approach friends of Grandfather's to borrow money. It must have been a very intimidating experience. One of Grandfather's friends told his servants not to open the door to Dad. Dad never forgave that person, and the dislike extended to the entire clan.[144] At the same time, Dad also made lifelong friends of those who were kind to him.

One person who really stood out was Arthur Morse, chief accountant of the Hong Kong and Shanghai Bank. When Dad went to see him, he immediately invited "Junior" (as he called Dad) to join him for lunch. He was extremely helpful, and Dad never forgot that.[145]

Grandmother received many offers to buy Grandfather's properties, but she, with Dad's encouragement, refused to part with any. In order to keep the properties, Dad had to make sure that money was available for mortgage payments. Mr. Morse made arrangements with Dad to have the Hong Kong and Shanghai Bank take over the mortgages with a monthly payment that was workable for the family. Grandmother and Dad had been afraid of Jardine Matheson repossessing East Point Hill, because the original mortgage was held by them.[146] The arrangements obviously worked well, because to this day, the Hong Kong and Shanghai Bank is still the family's bank in Hong Kong.

The fact that the Hong Kong and Shanghai Bank took over the mortgages turned out to be a blessing. When Hong Kong surrendered to the Japanese in 1941, there was complete chaos in the city. Our entire family went into China for the duration of the war, taking only what we could carry. Many of the deeds of our properties would have been lost during the looting if they had not been in the safe keeping of the bank.[147]

From the time of Grandfather's death, meeting the mortgage payments each month was the important event in the family. We were land-rich and cash-poor. Mok Guong was our rent collector and he made sure all our tenants paid their rents on time. Being the matriarch of the family, Grandmother was the person who doled out the housekeeping allowances for every household in the family, including Grandfather's siblings, despite the hard times we were experiencing.[148] There was never any dispute.

The entire family was very thankful that Grandmother refused to sell any of the properties. Over the years, real estate remained the main business of the family and led to the creation of the Hysan Development Company as a public company in 1981 with my father as the chairman.

Grandfather's spirit lives on by the good work of his descendants for the community and its educational institutions, and in the daily activities of the Hysan Development Company, now presided over by his eldest grandson, Lee Hon Chiu.

Bibliography

Barrie, Robin and Tricker, Gretchen. *Shares in Hong Kong*. Hong Kong, 1991.

Cameron, Nigel. *Power*. Hong Kong, 1982.

Chinn, Thomas W. *Bridging the Pacific*. San Francisco, 1989.

Correspondence between the Hong Kong government and the British Secretary of State for Foreign Affairs and the Secretary of State for the Colonies. 1927.

Cottrell, Leonard. *The Tiger of Ch'in*. London, 1962.

Endacott, G.B., and Hinton, A. *Fragrant Harbour.* Hong Kong, 1962.

Far East Practical English-Chinese Dictionary. Taipei, 1994.

Hong Kong Daily Press. February 29, 1928.

Hughes, B. O'Neill. *Companion to Chinese History.* New York, 1987.

Hughes, Richard. *Borrowed Place, Borrowed Time, Hong Kong and Its Many Faces.* London, 1976.

Jardine Matheson and Company, 1832-1932, Hong Kong. Privately printed, 1934.

Kadoorie, the Lord Lawrence, C.B.E., J.P., private letter, November 2, 1982.

Lee Family Ancestral Records. 河南龍利氏家譜。 Hand written in Chinese during the reign of Hum Fung (1851-62) in the Ch'ing dynasty, and updated to the present generation in print.

Lee Hysan Estate Company, Limited. *Memorandum and Articles of Association.* Hong Kong, November 20, 1956.

Lee Hysan Estate Company Limited private papers.

Lee Hysan Last Will and Testament.

Lee, Poh Kwan private letter, April 30, 1966, Penang.

Lee, R.C., C.B.E., J.P., LL.D., private letter, November 3, 1982.

Lowe, K.J.P. "Hong Kong, 26 January 1984: Hoisting the Flag, Revisited." *Journal of the Hong Kong Branch of the Royal Asiatic Society,* vol. 29, 1989.

St. Paul's Co-Educational College 75th Anniversary Commemorative Album. Hong Kong, 1991.

Seagrave, Sterling. *Dragon Lady.* New York, 1992.

Sinn, Elizabeth. "Kowloon Walled City: Its Origin and Early History." *Journal of the Hong Kong Branch of the Royal Asiatic Society,* vol. 27, 1987.

Siu Kwok-kin, Anthony. "The History of Hong Kong: From a Village to a City." *Journal of the Hong Kong Branch of the Royal Asiatic Society*, vol. 29, 1989.

South China Morning Post. March 1914. March, April and May 1928.

Stokes, Gwenneth and John. *Queen's College, Its History, 1862-1987.* Hong Kong, 1987.

Yu Sing Mau and Lau Cheun Foon. *Nineteenth Century Hong Kong.* Hong Kong, 1994. (Written in Chinese) 十九世紀的香港，余繩武、劉存寬主編。

Welsh, Frank. *A History of Hong Kong.* London, 1993.

Wu, K.C. *The Chinese Heritage.* New York, 1982.

Notes

1. The entire chapter is based on information from *The South China Morning Post,* April 29, 30, and May 1, 1928.
2. Lee Family Ancestral Records.河南龍利氏家譜。
3. Ibid.
4. Interview with Esther Lee.
5. Yu Sing Mau and Lau Cheun Foon, *Nineteenth Century Hong Kong,* Hong Kong, 1994. *十九世紀的香港*，余繩武、劉存寬主編， p. 235.
6. Chinn, Thomas W., *Bridging the Pacific,* San Francisco, 1989, pp. 3-4.
7. Interview with J.S. Lee.
8. *Bridging the Pacific,* p. 44.
9. Interview with Lee Pak Chao.
10. Interview with J.S. Lee.
11. Interview with Lee Pak Chao.
12. Siu Kwok-kin, Anthony, "The History of Hong Kong: From A Village to a City," *Journal of the Hong Kong Branch of the Royal Asiatic Society,* vol. 29, 1989. p. 391.
13. Welsh, Frank, *A History of Hong Kong*, London, 1993, p. 20.
14. Seagrave, Sterling, *Dragon Lady,* New York, 1992, p. 43.
15. Ibid., p. 43.
16. *A History of Hong Kong,* pp. 78-79.
17. Ibid., p. 79.
18. Ibid., pp. 76-77.
19. Ibid., pp. 76-77.
20. Ibid., p. 83.
21. Ibid., p. 84.
22. Ibid., pp. 84-85.
23. Ibid., p. 85.
24. Ibid., p. 87.
25. Ibid., p. 86.
26. Ibid., p. 89.
27. Ibid., p. 90.
28. *Dragon Lady,* p. 45
29. *A History of Hong Kong,* p. 93.
30. *Dragon Lady,* p. 45.
31. *A History of Hong Kong,* p. 94.
32. Ibid., p. 97.

33. *Dragon Lady,* p. 46.
34. *A History of Hong Kong,* p. 97.
35. *Dragon Lady,* p. 45.
36. *A History of Hong Kong,* p. 103.
37. *Dragon Lady,* p. 46.
38. *A History of Hong Kong*, p. 105
39. Ibid., p. 106.
40. Ibid., p. 104.
41. Lowe, K.J.P., "Hong Kong, 26 January 1984: Hoisting the Flag, Revisited," *Journal of the Hong Kong Branch of the Royal Asiatic Society,* vol. 29, 1989, p. 9.
42. Ibid., p. 11.
43. Ibid., p. 12.
44. *A History of Hong Kong,* p. 108.
45. Ibid., p. 105.
46. Ibid., p. 108.
47. Ibid., p. 110.
48. Ibid., p. 111.
49. Ibid., p. 124.
50. *Journal of the Hong Kong Branch of the Royal Asiatic Society,* vol. 29, p. 393.
51. *A History of Hong Kong,* p. 167.
52. Ibid., p. 197.
53. Sinn, Elizabeth, "Kowloon Walled City: Its Origin and Early History," *Journal of the Hong Kong Branch of the Royal Asiatic Society,* vol. 27, 1987, p. 30.
54. Ibid., p. 31.
55. Ibid., p. 31.
56. Ibid., p. 32.
57. Ibid., p. 35.
58. Ibid., p. 35.
59. Cameron, Nigel, *Power,* Hong Kong, 1982, pp. 14-15.
60. *Journal of the Hong Kong Branch of the Royal Asiatic Society,* vol. 27, p. 38.
61. *Journal of the Hong Kong Branch of the Royal Asiatic Society,* vol. 29, p. 391.
62. *A History of Hong Kong,* p. 279.
63. *Nineteenth Century Hong Kong. 十九世紀的香港，* p. 238.
64. Ibid., p. 165
65. *A History of Hong Kong,* p. 235.
66. *Nineteenth Century Hong Kong. 十九世紀的香港，*p. 336.
67. Stokes, Gwenneth & John, *Queen's College, Its History, 1862-1987, Hong Kong,* 1987, p. 6.
68. Ibid., pp. 4-6.
69. Ibid., p. 7.
70. Ibid., p. 27.
71. Ibid., p. 29.
72. Ibid., Foreword.
73. Ibid., p. 245.
74. Ibid., p. 15.

75. Ibid., p. 8.
76. Ibid., pp. 51-52.
77. Ibid., p. 249.
78. Ibid., p. 251.
79. Ibid., p. 98.
80. Ibid., p. 291.
81. Ibid., p. 98.
82. Ibid., p. 479.
83. *Nineteenth Century Hong Kong. 十九世紀的香港*， p. 370.
84. Ibid., p. 159.
85. Ibid., p. 371.
86. Ibid., p. 208.
87. Ibid., p. 330.
88. Interview with J.S. Lee.
89. Interview with Lee Siu Yuen.
90. Lee Hysan Last Will and Testament.
91. Interview with J.S. Lee.
92. *Queen's College, Its History, 1862-1987,* p. 6.
93. *South China Morning Post,* March 1914.
94. *Nineteenth Century Hong Kong. 十九世紀的香港*，p. 366.
95. *A History of Hong Kong,* p. 249.
96. *Power,* p. 79.
97. Ibid., p. 17.
98. Lee Hysan Last Will and Testament.
99. *Power,* p. 57.
100. *St. Paul's Co-Educational College 75th Anniversary Commemorative Album,* Hong Kong, 1991, p. 70.
101. Ibid., p. 73.
102. Lee Hysan Company private papers.
103. *Jardine Matheson and Company, 1832-1932,* p. 4.
104. Ibid., p. 5.
105. Ibid., p. 11.
106. Ibid., p. 11.
107. Ibid., p. 12.
108. Lee Hysan Company private papers.
109. Lee Hysan Company private papers.
110. Lee Hysan Company private papers.
111. Lee Hysan Company private papers.
112. Interview with Esther Lee.
113. Interview with J.S. Lee.
114. Lee Hysan Company private papers.
115. *Queen's College, Its History, 1862-1987,* p. 311.
116. Lee Hysan Company private papers.
117. Interview with Esther Lee.
118. Interview with Amy Lee Yuen.

119. Interview with Esther Lee.
120. Interview with Amy Lee Yuen.
121. Interview with Esther Lee.
122. Interview with Lee Pak Chao.
123. Interview with Hon Chiu Lee.
124. Related by R. C. Lee.
125. Interview with Hon Chiu Lee.
126. Interview with Lee Pak Chao.
127. Lee Poh Kwan private letter, April 30, 1966, Penang.
128. Interview with J.S. Lee.
129. Interview with Lee Shiu Yuen.
130. Interview with J. S. Lee.
131. Interview with Esther Lee.
132. Ibid.
133. Ibid.
134. Interview with J.S. Lee.
135. *Queen's College, Its History, 1862-1987,* p. 96.
136. Interview with Esther Lee.
137. Interview with Daniel Wong.
138. Interview with Jennie Hoo.
139. Interview with Esther Lee.
140. The entire chapter is based on information from *The South China Morning Post,* March, April and May, 1928.
141. Interview with Amy Lee Yuen.
142. The above information is based on *The South China Morning Post,* May 1928.
143. *Power*, p. 86.
144. Interview with Amy Lee Yuen.
145. Ibid.
146. Interview with J.S. Lee.
147. Ibid.
148. Ibid.

Map of China during the Spring and Autumn period (774-484 B.C.). The feudal kingdom of Zhau (in large square) was granted to our ancestor Hung Yik during the reign of the emperor Chou Sing Wang (1115-1078 B.C.)

春秋列國圖（公元前七七四年至四八四年），我祖熊繹於周成王時（公元前一一一五至一〇七八年）封彊於今日四川荊楚一帶。

City gate of Chu Gai Lai.
南雄珠璣古巷。

Great-grandparents' original home in the village of Hoiping.
曾祖父母於開平鄉下之住屋。

Garlieu Village.
嘉寮村。

Chicken coop built on stilts over the river in Garlieu Village.
嘉寮村溪畔雞棚。

Newborn piglets in Garlieu village. Pork is still very important in the Chinese diet.
嘉寮村豬舍；豬肉至今仍為國人主要食用肉類。

New ancestral hall built in Garlieu in 1896. Since the Communist Revolution, the building has been used for storage.
一八九六年所建之利氏祠堂，文化大革命後成為堆棧。

Front entrance of Great-grandparents' house in Sunwui city.
新會曾祖父母住屋之入口。

Drain pipe in the courtyard of Great-grandparents' house in Sunwui city.
曾祖父母於新會宅院屋頂之排水管。

Back garden of Great-grandparents' house in Sunwui city.
曾祖父母於新會房舍之後院。

Holland Yuen Road. The Lee family lived there during the years in Macao.
家人避居澳門期間所住的馬路名。

Grandfather.
祖父利希慎公。

Grandmother.
祖母利黃蘭芳女士。

Grandfather and my father.
祖父與父親。

1. Grandfather. 2. Grandmother. 3. Third Lady. 4. Second Lady. 5. Fourth Lady. 6. Fifth son Wing Kit. 7. Fourth daughter Joyce (Shun Kum). 8. Sixth daughter Amy (Shun Ho). 9. Third daughter Ansie (Shun Ying). 10. Second daughter Doris (Shun Wah). 11. Third son Harold (Hsiao Wo).

(1)祖父 (2)祖母 (3)三姨太 (4)二姨太 (5)四姨太 (6)五子榮傑 (7)四女舜琹 (8)六女舜豪 (9)三女舜英 (10)二女舜華 (11)三子孝和

From right to left: Grandfather, my father and relatives.
右起：祖父、父親及親人。

Grandparents in the garden.
祖父母於大屋花園。

Grandparents with the children in the garden.
From back to front: Left: fifth daughter Dione (Shun Yin), fourth daughter Joyce (Shun Kum) and sixth daughter Amy (Shun Ho).
Centre: fourth son Jung Sen.
Right: second son Ming Hop, fifth son Wing Kit.
祖父母與子女攝於大屋花園。
由後至前左起：五女舜賢、四女舜琹、六女舜豪。
中：四子榮森。
右：二子銘洽、五子榮傑。

A corner of the main hall of the Big House.
大屋客廳之一角。

View of Hong Kong from the main hall of the Big House at Chinese New Year. From left to right: fifth daughter Dione (Shun Yin), sixth daughter Amy (Shun Ho), fourth daughter Joyce (Shun Kum).
新年期間由大屋遠眺香港景觀。
左起：五女舜賢、六女舜豪、及四女舜琹。

Grandfather and Grandmother with the children at the main gate of the Big House.
大屋門前的祖父母與子女。

Grandfather and Grandmother with the children in one of the sitting rooms in the Big House.
祖父母與子女攝於大屋起居室之一。

Grandfather with the children on the steps of the main hall of the Big House. From left to right: third daughter Ansie (Shun Ying), second daughter Doris (Shun Wah), fifth son Wing Kit.
左起：三女舜英、二女舜華及五子榮傑於大屋台階前。

Fourth Lady with seventh daughter Diana (Shun Yee) on her lap and sixth son Jung Kong standing.
四姨太與七女舜儀、六子榮康。

The Big House.
大屋。

Lee Building.
利行。

Main gate of the Jardine Matheson estate in East Point.(1840s)
銅鑼灣怡和宅院之大門。

Photograph of East Point Hill in the early 1840s, showing No. 1 and No. 2 houses of the taipans of Jardine Matheson, with racing stables in the middle. Superimposed are the names of present-day structures.

一八四O年代東角（銅鑼灣）風貌，當時怡和大班住一號及二號建築物，中間為馬廄，箭頭所指為今日之建築。

Entrance to the Lee Gardens.
利園入口。

One of the taipans' houses in Lee Gardens.
利園內大班所住的華廈之一。

Lee Gardens.
利園。

Lee Gardens.
利園。

Another taipan's house in Lee Gardens.
利園內大班華宅之一。

Lee Gardens.
利園一景。

Lee Theatre photographed in the 1950s.
一九五〇年代之利舞台。

Original calligraphy done by Li Ching Yat for Lee Theatre which now adorns the main entrance of the Lee Theatre Plaza.

李精一先生當年為
利舞台劇院開幕
所寫之對聯，現
懸掛於新建之
利舞台廣場內。

Inside Lee Theatre.
利舞台劇院內景。

My father as a student in England.
英國求學時的父親。

My parents' wedding, photographed in front of St. John's Cathedral.
父母於香港聖約翰大教堂結婚。

My parents' wedding cake.
父母的結婚大蛋糕。

My parents during their honeymoon in Switzerland.
雙親於瑞士蜜月。

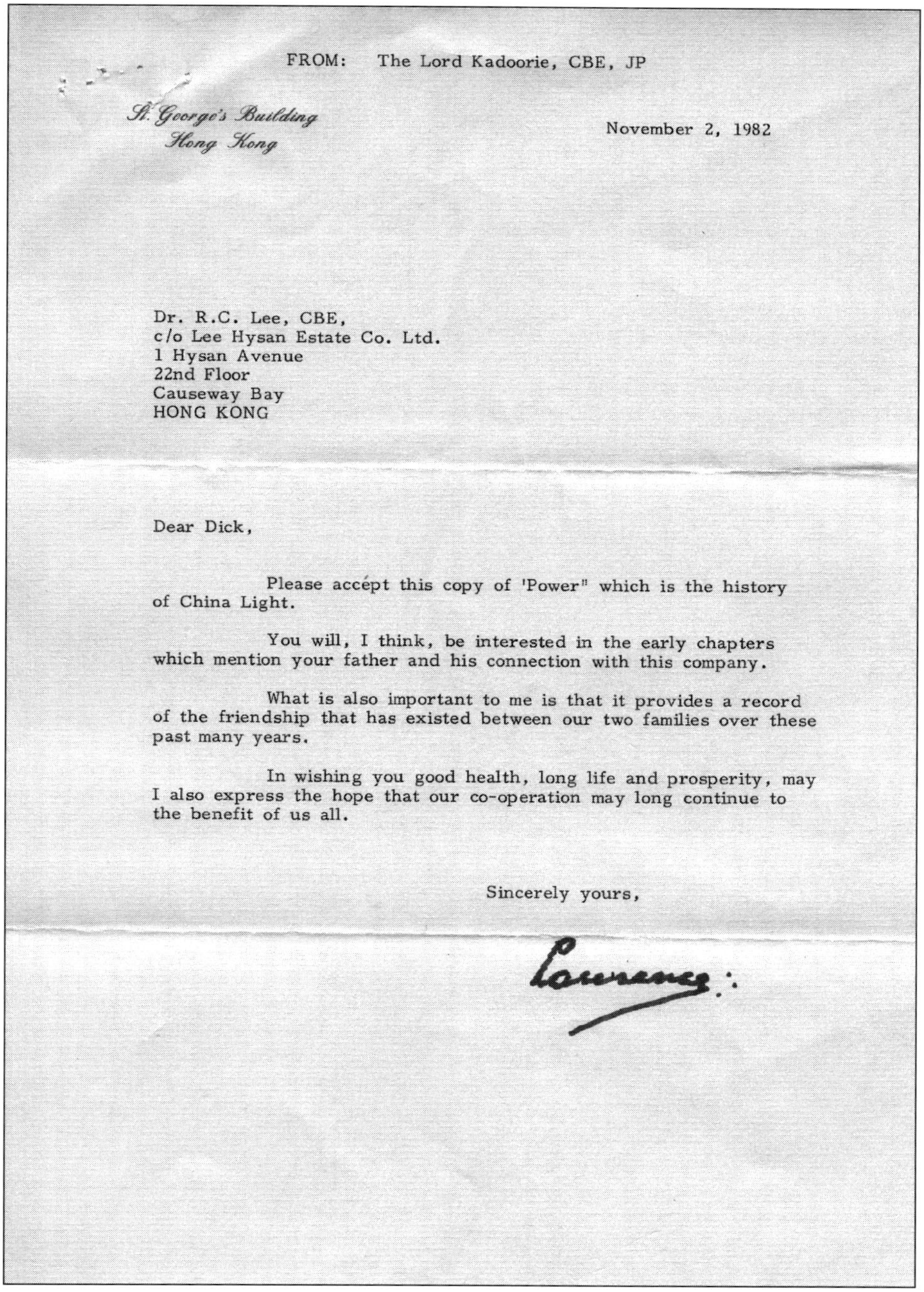

FROM: The Lord Kadoorie, CBE, JP

St. George's Building
Hong Kong

November 2, 1982

Dr. R.C. Lee, CBE,
c/o Lee Hysan Estate Co. Ltd.
1 Hysan Avenue
22nd Floor
Causeway Bay
HONG KONG

Dear Dick,

Please accept this copy of 'Power" which is the history of China Light.

You will, I think, be interested in the early chapters which mention your father and his connection with this company.

What is also important to me is that it provides a record of the friendship that has existed between our two families over these past many years.

In wishing you good health, long life and prosperity, may I also express the hope that our co-operation may long continue to the benefit of us all.

Sincerely yours,

Lawrence

Letter from Lord Kadoorie to my father in reference to Grandfather's position in China Light and Power Company.
*Lord Kadoorie*在致函父親時提及祖父與中華電力公司之淵源。

R. C. LEE CBE JP
ONE HYSAN AVENUE, 21ST FLOOR
HONG KONG
TEL. 5-775353

3rd November 1982

The Rt Hon the
Lord Kadoorie CBE JP
24/F St George's Building
Hong Kong

My dear Lawrence,

My grateful thanks for sending me a copy of "Power" which will be treasured by my family.

I knew something about my late father's connection with China Light in the early days as I was the one who had to arrange payment for his death duties.

Our two families are truly Hong Kong belongers and old friends. I feel sure that our efforts for Hong Kong will continue to contribute to the stability and prosperity for many more years to come.

With best wishes.

Yours ever,

Dick

My father's answer to Lord Kadoorie in reference to the sale of the shares of China Light and Power Company to Lord Kadoorie's father Ely.

父親回覆*Lord Kadoorie*信中亦提到當年他曾將中華電力公司股權售予給父親。

Wedding of second daughter Doris (Shun Wah) at the Lee Gardens. Taken in front of one of the taipan mansions. 1. Sixth son Jung Kong2. Seventh son Wing Tat. 3. Grandson Hon Chiu. 4. Third daughter Ansie (Shun Ying). 5.Groom, Cheang Goon Sing . 6. Bride, second daughter Doris (Shun Wah). 7.Granddaughter Helena (Hing Wan). 8. Eighth daughter Vivien (Shun Ngor). 9. Seventh daughter Diana (Shun Yee). 10. My mother. 11. Grandmother. 12. Aunt Tong Lai Wai, wife of Ming Hop. 13. My father. 14. Fourth Lady. 15. Third Lady.16. Second Lady. 17. Sixth daughter Amy (Shun Ho). 18. Fifth daughter Dione (Shun Yin). 19. Third son Harold (Hsiao Wo). 20. Second son Ming Hop. 21. Fifth son Wing Kit. 22. Fourth son Jung Sung. 23. Fourth daughter Joyce (Shun Kum).

二女舜華出閣當日家族成員攝於利園大班行之一門前。
(1)六子榮康(2)七子榮達 (3)漢釗孫(4)三女舜英 (5)新郎鄭觀成
(6)新娘二女舜華 (7)慶雲孫女(8)八女舜娥 (9)七女舜儀 (10)母親瑤璧
(11)祖母 (12)銘洽妻唐麗蕙 (13)父親銘澤 (14)四姨太吳佩珊
(15)三姨太蘇淑嫻 (16)二姨太張瑞蓮 (17)六女舜豪 (18)五女舜賢
(19)三子孝和(20)二子銘洽 (21)五子榮傑 (22)四子榮森 (23)四女舜琹

Author and third aunt Ansie (Shun Ying) being welcomed by the villagers in Garlieu in 1993.
作者與三姑姐舜英在一九九三年返嘉寮村時受村民歡迎。

Author with third aunt Ansie(ShunYing), paying respect to our ancestors, in front of our ancestral home in Garlieu village.
作者與三姑姐舜英在嘉寮村祠堂上香。

Author's and third aunt's family members in the sitting room of Great-grandparents' house in Sunwui city. Photographed during their visit in 1993.
作者與家人於一九九三年訪新會曾祖父母舊居。

Author with the children in Garlieu village in 1993.
作者於一九九三年返嘉寮村時與孩童合影。

Author's and third aunt's families in front of Leung Yik Hall, named after Great-grandfather. It forms part of the complex of the Sunwui City People's Hospital.
作者與舜英姑姐家人攝於新會市人民醫院之良奕堂前。

利休長流

利文長流

利德蕙著

CALYAN
PUBLISHING LTD.

利氏長流

謹以此書獻給

祖父 利希慎公之後裔

謝言

德蕙此書之得以順利完成，實有賴利家多位親人對我的支持，尤以四叔利榮森幫忙最大。他不只為我提供各種參考書籍、珍貴之相片及資料，更不厭其煩為我敘述往事；而漢釗及定昌堂兄亦為我搜尋了不少資料及相片；至於母親及舜豪、舜英姑姐，則告訴我許多陳年往事；英潔嬸母也非常熱心的為我搜集舊相片，我謹在此對他們致最誠懇的謝意。

同時我亦要一併多謝樹源叔公、百就叔公、寶祥叔父、錦彰舅父、舅母、瑤珠、瑤芬及瑤芝阿姨給我的幫助。

我的助手陳小芳不但為我將原稿譯成中文，並為我搜集了許多中文資料，我非常感謝她的幫忙。

另方面，我必須對多位中文學者，如顧漢昭先生、載國材先生、劉志平牧師以及香港大學及市政局圖書館工作人員的幫助致謝。

最後，我的丈夫伍衛權醫生，為此書做了不少攝影及複製舊相片工作；再加上大兒子偉雄給我數本珍貴的參考書，二子俊雄、三子廸雄及譚康樺小姐給我的意見，及二媳善婷為協助俊雄出版此書經常熬夜，我在此謹表衷心謝意。

利久長流

利德蕙著

楔子 - 暗殺

一九二八年四月三十日下午一時，香港威靈頓街裕記華人會所，有十多位會員正在用餐時，突然聽到連續三響鎗聲，跟著有人大喊「救命」，兩個會所的伙計急忙衝出查看。名叫羅流的伙計首先衝抵台階口，他沿著走廊跑去，看到遇害者扶持著牆壁，臉色蒼白，顯然已身受重傷。羅流瞥見一名身穿白袂短衫的男子，正由會所奔離，為了追趕這名男子，他無暇照料傷者，可惜一轉過街角這名男子即失去踪影，當會所的其他會友看到躺在走廊上的遇害者時，他已是回天乏術。

事發未幾，兩位印度警官及救護車相繼趕抵現場，將死者運至殮房。其後警察總監Hon. Mr. E.D.C. Wolfe、Murphy探長、Lane副探長及其他警員都先後趕到調查，並通知死者妻妾家屬親人前往認屍。死者身中三鎗、一鎗貫穿左肩、另一鎗擊中胃部，而第三鎗則射入他的後頸—這最後的一鎗顯然是在死者轉身時所發。其中一顆子彈更在穿過死者身體後，射進了走廊的水泥牆壁。

兇手的手鎗及三發空彈殼，事後被警方發現棄置於裕記會所的樓梯下。一九二八年五月一日星期二，南華早報大字標題報導此宗被稱為香港開埠以來最驚人的大案。死者為香港聞人富商，兇手於市中心光天化日之下行兇，顯見早有預謀。裕記會所會友指稱，眾人皆知死者為該會所定時午餐常客，也常在會所晚餐，是以兇手可以事先匿於暗處俟機行刺。據一會友指出，五、六年前也有另一羅姓富商在階梯處遇害，但事發地點則在會所入口處的水泥廻廊下。

裕記會所設於威靈頓街一幢樓宇的三樓，但大門在九如坊巷，平日少有路人，極為僻靜。威靈頓街與九如坊巷以一石板階梯小路接壤，直通會所大門。

本案行兇動機至今不明，警方得知死者生前曾多次受到威脅，並聲稱要在其長子的婚禮中投擲炸彈。其長子的婚禮僅在案發前不久舉行，當時此類威脅雖未曾成為事實，但其親友均勸戒死者勿去幽僻場所，死者卻顯然對勸喻不加理會而送命(二)。

楚公子

在孩提時代，當我在大屋祠堂祭祖時，我對神壇的祖先牌位充滿好奇迷惑。我原本一直以為我們來自香港，後來方知我們的家族源起於中國北部的河南省，繼遷廣東新會，再定居香港。

公元前一一二二年商亡周起，周武王克商之後，封其父為周文王，其時為中國社會最封建的時代。

先祖鬻熊公乃周文王之師，由於輔政有功，周成王時酬傭其孫熊繹贈為子爵，封疆在今之四川荊楚。後熊繹之子理真於河南利水一帶，採納田賦，遂以利為姓氏，此乃利氏得姓之由來。

綜觀我利氏家族，代出名人，成就不凡的大有人在。如在楚漢雙爭時代（公元前二二一年至二〇六年），先祖利璣即為西楚霸王項羽麾下名將。後劉邦滅秦為漢高祖，項羽烏江自刎，利璣為劉邦徵用，但其後背叛高祖，終為所擒，但得高祖恕免，且獲封地，告老還鄉。

漢朝（公元前二〇六年至公元二二一年）利乾為漢高祖劉邦之孫中山靖王丞相，利春林（正式記載一世祖）官居左丞相，利真源為東漢（公元二五年至二二一年）得道真人。晉（公元三一七年至四二〇年）則有利輅輔國愛民。

至唐時（公元六一八年至九〇七年）有利溉，名題雁塔，宿學鴻詞。唐貞觀六年（公元六三二年）定天下氏族共一百九十八家，汝南五門（周、袁、許、利、梅）及河南五門（華、蕭、陸、丘、利）均有利姓。

宋（公元九六〇年至一二七九年）有利申，立說行世，至明朝（公元一三六八年至一六四三年）利本堅職掌高權，治國平亂有功。

清朝（公元一六四四年至一九一一年）道光廿三年，利奕銓為科舉進士，官居要職，祠堂並

立碑留芳。(2)

由利理真時期至宋末，其間約二千五百年，利族已由河南利水南遷至廣東南雄府。

十三世紀中葉宋朝末期，即德照公之子利庚郎公時期，根據族譜所言，導致我族人由廣東省南雄府南遷至廣東省廣州府，有三種不同說法：

其一：載於「南雄來歷考」及「利氏河南、汝南二郡世系源流人物序」。南宋咸淳八年，宋懷王之蘇妃貌美性淫，至使懷王體弱失調，漸而失寵，被打入冷宮。

蘇氏設計喬扮村婦逃離京都臨安（今浙江杭縣），途中遇自南雄府運糧至京都返鄉的富商黃貯萬，蘇氏遂近船攀附，且示意願委托終身。貯萬慝於美色，便攜蘇氏返鄉，而不知此偽裝張姓村婦乃蘇妃正身。

日後聖上勅赦蘇妃，方知蘇氏已潛逃無踪。聖上大怒，四處通緝追捕不獲，後為貯萬家丁劉壯發現黃夫人之真正身份，洩密至京都。

聖上為挽回顏面，乃以南雄府有賊寇作亂為由，著派官兵掃盪，欲將全村居民斬殺。

南雄府珠璣里羅貴章女婿梁高輝任職京中，聞知此事，立刻密報回鄉，命鄉民盡速逃離南雄府。南雄府五十八村共九十七戶居民以南雄府謀生不易為由，具狀縣令欲南遷另謀生路，縣府批准限九年正月初十起程。全村除我庚郎公為朝廷舉人，及梁高輝兩家因任職京中尚無危險，未隨同遷徙外，其餘村民於是年四月十五日抵廣州府。

其二：載於「河南利氏族譜序」。南宋咸淳九年賈似道當權，一日皇上欲外出祭祀遇雨，便問大將軍胡顯祖當乘何匹馬，將軍建議騎「逍遙子」，議定之後請賈似道同行，賈似道因未得預問其事而不悅，便稱病不去。皇上為討好賈似道，遂將胡大將軍貶職，並令其妹胡妃削髮為尼。

後因皇后無子，乃立胡妃之子為太子。當時議曰天下豈有無母之君，遂須召胡妃回宮，才發

現胡妃匿居南雄府珠璣里，全村居民知悉後無不惶恐，便於咸淳九年集體請願遷居另謀生路，逃離南雄府。

其三：載於「河南利氏族譜序」。此傳聞亦提及賈似道，但說法略有不同。南宋咸淳九年八月聖上有祀于明堂值雨，聖上命車馬司官胡顯祖通知禮部權官賈似道同行，賈似道稱等天晴再去不晚。胡見賈意不肯同往，便偽覆聖上賈似道不欲同行。俟祭禮完畢，賈似道卻反稱身為禮部大臣，而皇上有祀於明堂乃不與其事，實不可留，奏懇請辭官職。聖上不明真相，追究原委，責怪胡顯祖覆命不實，遂將胡坐罪，其妹胡妃欲從中解救而觸怒皇上，遭逐出宮。

胡妃遭逐出宮後，為富商所收留，帶回南雄府珠璣里，詢問之下，方知其身為皇妃，立刻送離，不敢留宿，後為不良流民所辱，羞忿投水而死。珠璣里居民惟恐朝廷怪罪，遂集體申請遷移。

此三說不同姑備載於此。

一二八三年南雄之亂九年以後，宋亡元起，元太祖成吉思汗即位。時值禍亂之秋，庚郎公乃將家人遷至廣州府新會城與其他村民會合，在西花橋定居十年，後又遷往新寧縣。

庚郎公死後葬於新寧之名山—降冲，後因子孫未得祖墳風水遺福，竟將此山地賣於余姓人家，所幸利氏後裔再度將此福地買回。祖先對墓地風水之重視，曾一再於族譜中提示(3)。河南省利水早已乾涸，中國地圖已無此河名存在。

金山

曾祖父利良奕公于一八四一年清末道光二十二年生於廣東開平，而曾祖母則生於一八五〇年六月二十六日。兩人共育四子二女，居開平一個只有約二、三十戶居民之村落—水井坑。其時鄉間生活困苦，住處窄小溼暗。村民以種植稻米、菜蔬、甜橙、荔枝及飼養猪、雞、鴨為生，但由於土地貧瘠，農作收成僅足餬口。

窮困村婦生活尤為困苦，不只須洗衣煮飯，縫補衣物，養育子女，亦須下田做活，甚至分娩當日亦須照常工作。若無暇回家坐褥，便在田邊就地產兒，手邊如無刀具，就用瓷片鋒口清洗割斷臍帶，然後以布包裹新生嬰兒，置於一旁，又立刻繼續農活[4]。若無天災，村民尚可勉強維生。

十九世紀末，滿清政府腐敗無能，受盡歐洲國家欺凌，迫使清廷對貿易讓步。加上戰亂不斷，乾旱、飢荒相繼而至，民不聊生。中國南方尤為艱苦，許多壯丁以「賣豬仔」方式賣到東印度群島，馬來亞，印度洋之西印度群島或澳洲工作。他們被裝上髒亂污穢、擁擠不堪，有「水上地獄」之稱的船隻遠渡重洋。在途中經常發生暴亂，船長及船員常被殺害。在運送過程中不少壯丁死於飢餓或疾病，死亡率最高曾達六成多[5]。一旦到達目的地後，他們仍要在惡劣環境下做卑微粗重工作，並常受肉體折磨。

長久以來政府一直嚴禁人口外移，但始終無法禁絕。在一八四九年及一九五一年，加州及澳洲先後發現金礦，吸引大批外來淘金人潮。

一八九〇年曾祖父利良奕公由廣東遠赴又名金山的加州三藩市謀生，盼能寄錢返鄉養家活口。我想當年四十九歲的曾祖父是滿懷淘金美夢起程的，但顯然美夢難圓，因為除非礦坑已被發掘淨盡，中國礦工根本被禁止在礦藏豐富的礦坑工作。中國工人僅能做些苦役，幫廚等低微的工作。部份人夢醒絕望之餘只有返鄉一途，而曾祖父則在位於三藩市DUPONT街十七號的三藩市聯

合煙草公司覓得工作。該公司曾於一八七六年名列三藩市華人工商名錄內。

曾祖父定居三藩市後，遣其弟文奕返鄉將次子廷羨帶往，文奕公亦將其長子帶同隨行。

祖父利廷羨公—後人稱利希慎，一八八一年九月二十日生于廣東省開平縣，赴美與曾祖父團聚時年僅九或十歲。至於曾祖父當時為何指派次子而非長子赴美令人費解，或許須長子留鄉照顧曾祖母，亦或曾祖父已洞察其次子的潛力。

早期舊金山華人社團各自獨立，外人對留辮身穿寬鬆布衣鞋袦的中國人充滿好奇。十七世紀明亡清立，滿洲人統領中國，滿清為貶低漢人，令漢人剃額留辮狀似馬尾，並穿着形如馬蹄的布履。

中國早期移民鄉土觀念極重，都希望儲夠錢財後衣錦還鄉。初踏足新大陸者，皆仰賴親人或社團協助立足，在中國人初抵加州時即有此類社團成立，他們設立中文學校，並領導長期反抗歧視。中國的節慶習俗、地方戲劇更成為早期海外僑民的精神支柱。如遇有糾紛，均由社團或華人統一慈善會處理解決，各類黨派常取代警力提供保護。

舊金山之早期華人全來自廣東省，華人統一慈善會又稱為中華六會館，監管華人社區各類活動。一八四九年首先成立岡州會館，由廣東省六個不同鄉鎮成員組成。一八五〇年南海、番禺、順德三處華僑組成三邑會，一八五二年至一八五四年相繼有香山，客家本地華人各自組成會館(6)。曾祖父與其弟均屬四邑會會員，由於早期華僑均說家鄉方言，任何同鄉音者均視為兄弟。

利氏居舊金山期間，祖父廷羨公與堂弟就讀於Clay街九二九號的華人小學(7)。舊金山公立小學教育局為將白人與中國人學童分隔教育而設此校。一九一五年此校更名為東方學校，一九二四年再改稱為Commodore Stockton School(8)。

多年來唐人街居民只能在當地教會所提供的英語課程學習英文，祖父當時於舊金山所學的英文，對其日後發展助益極大。

一八九六年曾祖父良奕公與弟弟及兩名兒子，帶着多年金山工作的積蓄返鄉。曾祖母及家人居開平鄉間，生活困苦，見鄰近新會銀州湖四周常受湖水泛濫，土地肥沃，狩獵農耕均宜[9]，曾祖父及祖父返鄉後便舉家遷移至新會嘉寮坊雙水鎮。由此時開始，新會成為利氏宗祠所在。

曾祖父攜着積蓄由金山返國後，於嘉寮坊購置一大片土地築屋與曾祖母安享晚年，利氏太祖祠堂亦於此時興建完成[10]。文化大革命期間祠堂不幸全毀，磚石均被拆遷另建[11]。祠堂舊址近日新蓋之建築，有廣濶花園圍繞，遍植菓樹。利百就叔父去嘉寮坊時曾在此屋小住。

小島

曾祖父返鄉退休後生活頗為愜意，但為着子女有更好前途起見，於一八九七年將次子廷羨——我的祖父，送往香港皇仁書院就讀，由於曾祖父經常出門在外，故對香港十分熟悉。香港為當時華人出國之主要港口，當曾祖父渡洋赴舊金山時或許便是由香港登船。日後曾祖父由金山匯款回鄉亦經香港轉撥，他對中國商人在英殖民地發跡亦時有所聞。

以絕佳港口聞名於世之香港，歷史正式記載始於明朝，為中國南方一海島村名，島上盛產製莞香的香樹。明朝時香木及其製品多在石排灣聚集，再運至廣東，江蘇，浙江等地，因此該村稱為香港村，即今日香港之名由來。

明朝時島上經濟、人口不斷成長。萬曆年間島上至少有七個村落沿南岸而居，香港村乃其中之一，北岸則人口疏落。明末香港島雖受船艦巡守保護，但仍常遭海盜侵犯。一六六一年清初康熙元年，因流寇作亂，清廷遂勅令島民棄島遷居內陸。八年之後遷居勅令解除，村民始陸續返島重建家園。雍正初期已有：香港，鉄坑，舂坎，赤柱，大潭，筲箕灣及黃泥涌等七個村落重建完成，政府並於沿海設海防基地抵禦海盜，但島上並無任何軍防基地。

嘉慶年間（一七九六年至一八二〇年）大鵬營所駐守之紅香爐海防基地[12]，及其他兩個村落亦於此時完成。在一八四一年，此人煙稀少、荒蕪、多石的小島被割讓予英國。

十九世紀末，利氏家族開始定居於香港。利氏遷徙香港的原因乃因此小島是英國殖民地，而當時中國南方情況混亂，若香港仍隸屬中國，則情況必無改善。至於香港如何淪為英國殖民地，我必須回顧歷史。

鴉片問題

十七、十八世紀時期，西方對中國的產物如茶，絲，瓷器等需求日增，由是中國與美、英、法、荷蘭、葡萄牙等國間的貿易逐漸增加，但仍受朝廷諸多限制。朝廷認為中國乃天朝，所產物資僅供本國之需，毋須供應所謂「洋鬼子」。外國商人如要與中國交易須用紋銀。

由於英人嗜茶，英國遂於十八世紀末成為中國主要貿易國家。一六六四年時僅有兩磅茶葉進口英國，至一七八三年時，合法輸入英國之茶葉激增至二千六百噸，但估計尚有約三倍此數字的茶葉經走私途徑入英，以逃避關稅(13)。

當時貿易只限於廣州進行，中國其他港口均不對洋人開放。交易的貨品除茶外，尚包括絲、棉、瓷器、紙張、醫藥及香料，全以現金進行。由於受季候風影響，貿易季節僅限於五月至十一月。在此期間洋人才可於廣州市郊的廠房居留，外商將其商號設於下游葡萄牙管轄之澳門(14)。清廷非但禁止外商學習中文、僱傭、坐轎、持械，亦不許家眷同行，但以上限制並未強力執行。

當時對外貿易受南海海關之關津嚴格控制。關津之官職遠低於兩廣總督及省長，但後者俸祿低微，須仰賴關津不時進獻以供享受。理論上中國海關之職責雖有明確規定，但實際上，官員對外盡量壓榨歛財，對北京則盡量隱瞞少報。

由於外商僅可與關津轄下之商人交易，而此批商人又時受關津壓榨，故須千方百計對外商提高貨價以謀取重利，因此令外商議價極為不易。

各外國使館均派員至北京交涉，以謀求更好的官商關係，但因朝廷的民族優越感，不屑與外人談判而失敗。

儘管中西貿易受到諸多制肘，但一直續有成長，可是西方政府對庫存白銀不斷流往中土甚為憂慮，結果發現可用鴉片與中國交易讓白銀回流。

中國初時種植鴉片只為醫療之用，但十七世紀時國人漸漸吸食上癮，政府為控制鴉片泛濫，在一七二九年開始對鴉片徵課重稅，並由政府專賣，壟斷市場，成為權貴之專利品(15)。至一八〇〇年鴉片被禁止入口，而英商眼見廉價印度鴉片在中國中、下階層市場大有可為，於是便勾結華商，行賄官員，大量走私印度鴉片進入中國，使吸食鴉片的人口大增，甚至宮中太監亦多上癮。此後數以百萬磅計的鴉片經合法或走私途徑源源流入中國。全國茶葉的銷量都無法抵償此數，朝廷庫存銀兩日漸流失，雖然政府立法嚴禁吸食，但百姓仍視若無睹。

十九世紀中葉，正值英國國力全盛時期，而滿清政府自一六四四年統治中國以來，當時正為最積弱的年代，貪風熾烈使腐敗的政府官員直接參與鴉片貿易，而鴉片貿易亦演變成中國主權的鬥爭。洋人常各方挑釁華人，倘遇華人反擊，洋人即向當地官吏求償，若不得逞，即招炮艇入港示威。洋人從不認為懦弱的滿清政府，會採取強硬政策與之抗衡。

輸入中國鴉片之數量快速增長令人担憂，在一八三五年時已達三萬箱。當時鴉片貿易，除須支付中間人及中飽私囊之官吏強索外，均為非法進口，無需納稅，加上所有非法進口鴉片皆以金銀交易，令國庫損失甚鉅。在一八三六年北京政府曾對鴉片進口應否予以合法化開放，抑或全面禁止而爭辯。

廣州商人則一致要求應採開放合法化的政策。據傳聞所說，皇后亦同意開放政策，而一姓徐之資深法官亦對此事撰文發表，其英文譯稿亦在一八三六年七月十二日出版之「廣州記錄報」登載(16)。廣州之華洋商人皆視鴉片合法化為理所當然，但怡和 (Jardine Matheson)英商 James Matheson 則持異議。他在七月十二日寫到「雖然目前鴉片價格已經提高，但對我方利益來說，本人並不看好(17)。」經過了半年之久，清廷仍毫無動靜，廣州的商人方明白原來道光皇帝非但不許鴉片合法化，並決定全面禁煙，唯一問題是如何中止鴉片貿易。

十八世紀初與中國進行貿易的兩家主要英商，分別是 Dent and Co., 及Magniac and Co.。Magniac and Co. 後來由 Dr. William Jardine 及 James Matheson 組成之 Jardine Matheson and Co. 取代。這些英國公司俗稱為洋行，合伙人則稱大班，Jardine Matheson 公司為怡和洋行。Dent's 即寶信行。

一八三六年James Matheson及William Jardine返英鼓勵英政府對華採取強硬政策，並建議若中國政府拒絕英商之要求，則英政府應考慮侵佔舟山群島，當時香港根本不在考慮之內。Matheson在所著「中英貿易之今日及其展望」一書中，將中國人形容為極其低能愚蠢、貪婪、頑固、狂妄自大的民族，並對無辜的外國人冠上各種惡名，肆意欺凌，更堅稱大英政府須為國雪恥，保障商業利益(18)。

當時指揮官Charles Elliot對英外交部長Palmerston稱，英人對華人侮辱之行為是華人反抗主要原因。不幸Elliot無力處置或懲罰此等人物，而東印度公司亦堅稱中國人是偉大而獨特的民族，決不會接受任何條約的脅制。在無任何協約之下，英國不能漠視國際公法而壓迫中國(19)。

在華美商警覺此種劍拔弩張情況，當時美國最大之洋行Russell's順勢決定對中國終止此項風險日高，惡名昭彰的鴉片貿易(20)。

至於甘冒風險之英商，則已做最壞打算，準備到時求助於英國皇家海軍，但怡和洋行為謹慎起見，乃主張將鴉片及其他貨品全部經由香港轉運至中國沿海地區，由此可見當時香港水域已有商業活動(21)。

江蘇省省長林則徐於一八三九年三月受朝廷指派為欽差大臣，全權處理廣州鴉片問題，並以兩湖（湖南、湖北）總督身份下令：凡任何煙民拒戒煙者，將被割去上唇，使其永不能再使用煙槍。

林則徐明白表示其懲治對象為華人不法份子，對遵守中國法令、不走私販毒的外國人必妥善

照顧，並重申合法貿易之重要及必予保護。三月十八日林則徐宣佈將沒收所有庫存鴉片，外國商人並須立切結書保証永不販煙，若有不從者，華商可判監禁下獄，財產充公，甚至斬首；外商則將依新訂律法嚴懲，並限期三日繳出鴉片充公，此三日之內不許洋人離開其廠房範圍[22]。

在華外商經由三月二十一日會議商定，同意僅繳出為數一千箱之少量鴉片，林則徐對此敷衍態度大為震怒。由於 William Jardine 仍在英國，林則徐便下令拘捕寶信行的鴉片走私頭子 Lancelot Dent[23]。當時英國指揮官 Elliot 欲將 Dent 偷渡出境，但中國官方已在河道設障阻擋。但 Dent 最後仍然脫逃至馬尼拉，購屋另設鴉片銷售總部。

英政府代表Elliot指揮官指示英商繳出鴉片，並表示英政府事後會負責補償其損失。實則英政府根本無意貼補英商損失，除非可以施壓計誘中國政府補償費用。結果有二百六十多萬磅的鴉片繳集在珠江河畔之特定地區全數銷毀[24]。

怡和洋行負責人James Matheson表面上同意停止進口鴉片，他深知居時只須辯稱切結書為被迫所簽，所以不具效力。他並去信鴉片供應商，將鴉片轉運至澳門的Alexander Matheson繼續代銷[25]。

中印倫敦協會在一八三九年十一月二日對外交部長Lord Palmerston所詢意見回信中詳細指出，協會會接受英人在中國須遵守中國律法及鴉片禁運之事實。信內稱：「我方必須承認只要中國堅持禁止鴉片進口，英國商人就必須遵守中國律法。若有任何英人觸犯中國律法，大英政府不得干涉及予保護。」協會並認為英方代表絕對需要直接與中國政府談判，在同等條件下開放廣州以外的港口通商。若以上條件不為中方接受，則建議可採「以購買或其他方式取得一島嶼作領土」。Palmerston完全接受以上建議[26]。

以民權黨議員J. A. Smith為首的代表團包括Alexander Matheson及William Jardine亦對上述備忘錄加以贊同。J. A. Smith亦是怡和洋行的銀行代表。William Jardine其後不久成為Ashburton區

民權黨議員。該個代表團雖然無法說服Palmerston在政府賠償方面做任何讓步，但卻令Palmerston接受其可以不難強迫中國政府賠償煙款的說法。代表團並建議實際辦法，英內閣亦同意Palmerston立刻派員前赴中國(27)。

林則徐下令所有外國商人簽下保証書後，才准許外商所屬之商船重返黃埔。（黃埔位於廣州南方約七哩，為船舶進入珠江內陸的最終點。）所有丹麥、德國、美國及西班牙的商船船長在簽妥切結書之後，均得自由進出黃埔載卸貨物，唯獨英指揮官Elliot不服，憤而撕毀保証書，更隨即宣佈所有英國船隻貨物在澳門卸貨(28)。林則徐繼而下令禁止英人在澳門經商，英商便直接勾結中國走私販在海上進行交易。

英殖民地

一八三九年七月七日九龍尖沙咀英船水手酒醉滋事，一名英國水兵將華人平民林維喜殺害，並傷及其他數人[29]。Elliot欲掩蓋事實真相，僅付傷者少許慰問金，及對死者家屬象徵性賠償以換取一紙意外死亡之聲明，並經由預先安排之所謂法庭公審，宣判兇手無罪。林則徐堅決要求英方交出兇手，但為Elliot所拒，並將行兇水兵及其同伴送返英國釋放[30]。

八月十五日中國政府與澳門葡萄牙人聯手將居住澳門之英人驅逐出境，全體英人登上約六十餘艘船隻駛向香港停泊。清廷敕令禁止九龍村民對英國民船供應飲水及食物，但Elliot卻游說九龍當局網開一面。九月四日有數艘村民船隻載滿英人補給之物駛向香港時，被清朝水師船攔截，Elliot自Louisa號軍艦開炮反擊，由此點燃了戰端[31]。

林則徐究竟要將香港水面之英人船隻扣留多久，引起不少揣測。事實上若英人接受鴉片禁運條件，他便會在九月准許英人重回廣州繼續其正常貿易，但在Elliot禁止英商接受條件情況下，林則徐做出摧毀所有英船之威脅。十月二十六日林則徐對Elliot下最後通牒後，Elliot立刻警告所有英國船隻速離香港。經數度談判失敗後，一八三九年十一月三日英Volage號開火擊沉四艘清朝水師船船，於是揭開戰幔[32]。

在倫敦，國會的反對派保守黨議員William Gladstone對此大肆抨擊稱：「我從未見過比此次更加不仁不義，令國家蒙羞的戰爭，我大英帝國的旗幟竟用於保護此等可恥行徑，……我方應速退出[33]。」

在中國，林則徐下令全面禁止英人入境廣州之後，仍留在當地的美國商人立刻取代英人貿易角色。合法英國貨品仍可正常進口，至於鴉片仍繼續在沿海地區銷售，馬尼拉則成為鴉片運送中國東岸之轉運站。怡和洋行在鴉片遭沒收後所受之損失迅速恢復[34]，外交大臣Lord Palmerston在

致Elliot函中告知，由其堂兄George Elliot司令官所指揮的一隊海軍艦隊及一小隊陸軍，將於次年三月抵港。

不幸清廷對倫敦官商間微妙關係全不了解，亦低估英海軍力量。英國外交部發動戰爭的另一動機，是由於滿清政府一直拒絕與英國直接談判，英人欲藉此戰迫使北京清廷就範(35)。

清廷為和緩緊張局勢避免戰爭，另指派世襲高官的直隸（今湖北）總督琦善取代林則徐職位(36)。出於清廷意料之外，英軍並不強攻廣州，反而沿海直上撲擊毫無防禦準備的揚子江口之定海。當時約有上萬悍鍊英軍，對全無對外作戰經驗，亦未見過新式大炮的清軍展開攻擊。廣州寧波及其他港口亦續有戰事發生，直至北京俯首妥協(37)。

一八四〇年二月二十日Palmerston致清朝廷宰相書中指出：「英國政府要求中國割讓沿海一個或一個以上面積廣大的島嶼歸由英國永久管轄。」G. Elliot司令官及C. Elliot指揮官奉派為英國全權代表處理此事。此事在司令官G. Elliot返英之後便全權交由指揮官C. Elliot負責。英方原本屬意北面的舟山群島，但為中方所拒。

Lord Palmerston曾將寧波港納入妥協條件之內，而William Jardine則建議要求清廷割讓福摩沙（今台灣）、金門及廈門。當時僅有幾個村莊的香港則從未提及(38)。

一八四一年一月十五日琦善致書C. Elliot謂僅可在香港或九龍選擇其一，C. Elliot次日覆書接受香港。

身為職業海軍軍官的C. Elliot選擇香港的原因，是因香港具備優良海港各方面的條件。當時海軍大臣Sir John Barrow形容鯉魚門（海港舊名）是一個適合任何噸位船隻的絕佳海港，戰時利於防守(39)。島上水源豐富，僅須數台大炮，少數兵力及一艘戰船即可保護往來商船。另一方面台灣面積大過愛爾蘭，或許較難佔領控制。

一八四一年一月二十八日川鼻會議中，英方代表Elliot指揮官提出將香港正式割讓予大英帝國，但香港稅收仍歸中國。會議中中國並簽訂以分期付款方式，分六次共賠償英國六百萬（西班牙銀元），建立直接平等外交關係，及開放廣州對外貿易[40]。

英海軍艦隊在司令官Gordon Bremer爵士統領下，於一八四一年一月二十六日星期二上午八時，在太平山下據點「普慶坊」正式登陸佔領香港[41]。八時十五分大英國旗在普慶坊冉冉升起，官兵並全體舉杯慶賀，三呼女王萬歲。

英國佔領香港之後，首得其利的為商人及傳教士[42]。他們深知英國在中國南方擁有殖民地為他們所帶來的好處。當天有多位商人出席升旗典禮，其中包括至少四名印度商人，怡和洋行的James Matheson亦自澳門來港參予盛會，事後並繞島一周。二月八日有八名新教傳教士在澳門包下一小船去香港一探究竟，他們在勘查之後斷言：「香港在英國擁有主權的情況下，其影響力及重要性必然提高，並將成為東方領海上第一個重要商業島嶼[43]。」

維多利亞女王在去信其叔父比利時國王Leopold中說：「Albert對我擁有香港島覺得真是有趣，他認為維多利亞除了稱皇家公主之外，也應稱為香港公主[44]。」

Elliot指揮官在H. M. S. Wellesley號船艦上正式宣佈香港現已屬於女王陛下統治，目前暫由其本人代管[45]。在二月一日英方並對中國島民宣佈香港已割與大英帝國，但戰事仍繼續發生。

在商議割取港島之時，Charles Elliot對英政府的指示全置之不理，一八四一年四月二十一日Lord Palmerston在Charles Elliot免職信中稱：「閣下割取香港此等不毛之地的荒島，除可提供英居民在島上築屋退休之外，此島顯然將無法取代現今廣州之貿易地位[46]。」

清政府對琦善在川鼻條約中與英國所達成之協議全予否認，並將琦善捆綁押返北京，另派楊芳[47]取代其職務。一八四一年五月二十一日清廷派遣增援部隊赴港以收復失地，對停泊港內的英艦

隊開炮攻擊，但終不敵火力強大的英方H.M.S. Nemesis號軍艦，英軍並立刻受命直逼廣州[48]。

戰事一直持續至一八四二年的七月，清廷被迫於一八四二年八月二十九日在英艦H. M. S. Cornwallis號簽訂南京條約，此為中國歷史之轉捩點。條約中訂明中國對外開放廣州、上海、廈門、福州及寧波五個新商埠；對林則徐沒收燒毀的鴉片，須賠償六百萬西班牙銀元；償還中國商人所欠英商債務三百萬元；並賠償英國對這次戰爭的軍費一千二百萬元；以及將香港割讓給英國。不過要到一八四三年一月四日，新任外交大臣Lord Aberdeen（取代Palmerston職位）才指派H. Pottinger爵士出任香港總督[49]。

一八四三年六月二十六日，滿清政府終於承認南京條約，將香港割讓予英國為殖民地。

香港早期景氣凋零，根據一八四一年五月十五日人口調查報告，島上僅有十六個村莊，人口七千四百五十人[50]。一八四五年八月二十九日英殖民大臣接獲報告指稱：「香港全無商業氣息，僅是個英政府人員及商人居住的人煙稀少的地方[51]。」

一八五一年三月八日經濟報稱，香港是個山陵屏障地形絕佳的港口，島上僅有怡和洋行及寶信行兩家有成就的英商繼續其原有貿易業務。但報導中又稱香港是與中方貿易之庇護港，如能充份利用此殖民地，各方面均可有所改觀[52]。

香港隣近中國大陸本土，自始即令英軍擔憂。雖然九龍半島僅有少數以採石及燒石灰為生之村民居住，但半島西北角的九龍城寨於一六六八年即為第一個設有信號站的要塞[53]，其中的炮台極易攻打香港。

清政府對九龍戰略位置十分了解，一八四三年南京條約簽訂後，朝廷指派新安縣令赴九龍負責管轄九龍半島上的四百九十一個村莊，又另任大鵬營副將掌管軍務，並增兵至一百五十人駐守[54]。

清廷很快發現以上措施仍屬不足，一八四六年兩廣總督（廣東、廣西）耆英建議朝廷，應重

視九龍突顯之地理位置，認為亟須築牆護城，設置炮台以加強防禦措施；並應興建衙門、營房為軍隊提供住處（自始則借住民宅）及訓練場所。耆英認為以上策略不但可對駐香港之蠻夷產生「阻嚇作用」，亦可大大加強該區海防力量，是以九龍城寨於一八四七年興建完成(55)。新安縣縣令在「龍津義學」奠基典禮致辭時表示：九龍居民漸受港地蠻夷影響，世風日下，確有捐建民間義學，以振興九龍風氣之需要，更希能藉此感化外夷，以柔和其獷悍之氣(56)。

九龍及香港中英政府雙方官員都為維護當地治安而保持密切連繫。

一八六〇年中英北京條約簽訂，九龍半島讓予英國，但九龍城寨仍屬中國。中國官員仍繼續參與國界雙方的民間活動，許多廟宇均刻碑紀念當年各種事蹟(57)。

英屬香港島上華人居民遇有困難時，常求助於對岸九龍城中國政府。如一八八六年間謠傳大潭儲水工程須獻祭五百名孩童，結果當地兒童全被送至九龍接受保護，兩日之內全城不見任何孩童踪影(58)。

一八九八年七月一日深圳河以南之新界地區被英國強行租借九十九年，當地居民對政府避不協商，即將他們世居家園土地移交英人管轄之舉動，均感憤慨。在英國接管新界升旗典禮的前夕，港警及護衛隊在預定旗桿豎立處約八百米的山坡上，發現一個藏滿舊式武器的戰壕，欲用來埋伏襲擊參加升旗典禮之來賓及官員。政府派軍至山坡追捕鄉民，搜獲大批舊式武器，足可充當一兵工廠，結果升旗儀式戶外慶祝活動被迫取消(59)。

一八九九年清政府被迫自九龍城撤退(60)，當今英屬殖民地之香港，是香港島、九龍半島、新界及其附近二百三十五個衛星島嶼之總稱(61)。

一八八九年Rudyard Kipling問到香港：「為什麼這裡的人身上都可以嗅到一股銅腥味？」有人答說：「因為這個島景氣強勁，什麼都講錢」，此詮釋到今日仍為代表香港的寫照(62)。

實習教師

香港被割讓予英國之後，華人雖備受歧視，但仍成為大陸華人之避難所。最初僅有罪犯及亡命之徒潛逃來此，直到太平天國動亂期，（一八五一年至一八六五年）才成為富有體面之華人理想居所。又因香港為自由港沒有戰亂，許多中國商人亦攜眷長居此地。

他們的經驗及企業精神帶動了市面繁榮，到一八五五年時香港已不須英國政府補助，並成為中國南方主要貿易港。中國四分之一進口貨物及三分之一出口，皆由香港進出(63)。一八五九年時共有六十五家具有規模的華商註冊登記為行號。

自一八六一年起香港漸成為貿易及製造業的經濟中心，包括製糖、水泥、紡織、造紙等工廠，並設有銀行及保險公司為商家提供貸款服務，使小型企業有能力與大公司競爭。儘管華人在港仍受二等公民之歧視，但在商場上均十分成功。

當時財力最雄厚的英國大商行以怡和洋行為首，其成員不但在英國議會深具影響力，亦可左右香港總督之政策。他們在行政局、立法局均佔有議席，並控制了警察。該公司成員也多為最高級的香港會及馬會會員(64)。

一八六六年世界金融受倫敦Overend and Gurney跨台影響極大。中國沿海的怡和洋行靠變賣資產及出租其大綜產業而逃過破產，至於寶信行則關門大吉。十五年後香港前十八名納稅公司中，僅有怡和洋行是外籍公司，其餘皆為華人(65)。一八八〇年間華人已成為香港殖民地最大房地產業主，投資地產金額亦高出歐人七倍半(66)。儘管華人在港毫無政治權力，鮮受法律保障，又受盡歧視凌辱，但仍認為香港是個做生意的好地方。

由一八四〇年開始，馬禮遜教育會、羅馬天主教會、英國聖公會及倫敦傳教會開始在香港創辦學校。這些神學院主要目的是為其教會培養傳教士，但可負擔學費的華人中等家庭對西方教育

不感興趣，所以這些學校一無所成(67)。

James Legge牧師（後稱博士），是倫敦傳教會長老傳教士，為倫敦牛津大學第一位中文教授，能譯中文古文。他在一八五八年赴廣州參觀科舉考場，對省都會試考場內的學員們印象深刻，他認為中國是世界上對學術成就最為尊崇的國家(68)。

一八五〇年代香港人口繼續增長，但神學院卻漸減少，原因在於華人對英國人不信任，以及中英戰爭引起的敵意。Legge博士認為如果要獲得華人的支持，惟有將學校非宗教化，讓華人接納以吸引他們接受西方教育(69)。

Legge博士鑑於華人對英國之敵意及教會學校之失敗，於一八六一年創辦中央書院，將所有公立學校由一個「大中央學院」統轄管理，並提名Frederick Stewart（後稱博士）為首任校長。Legge博士深信在一八五〇年間移居香港之華人家庭，很快即能發現學習英文在商場的方便。校方揉和了中國的傳統知識及西方的自由思想，並兼顧華人對未來的希望。在這裡，不同種族男孩共同學習，相處融洽。校方也多次受到北京朝廷遣派特使表揚其學生在京城文官會試的優異成績(70)。

至一八六九年，一批英語流利又熟悉西方政治經濟型態的香港華人開始加入社會。

中央書院在一八八一年改稱維多利亞書院，一八九四年再改為皇仁書院。學校最引以為榮的是國父孫中山先生亦是皇仁書院的校友。先生於一八八四年四月十五日以孫帝象之名登記入學，編號二七四六(71)。英殖民香港皇仁書院水準與國內名校Harrow及Eton齊名，校譽之高世界著名。校友中不乏社會菁英人士，祖父即是其中之一(72)。

曾祖父相信其家族在香港可以有所作為，便將次子送入皇仁書院就讀。一八九七年為維多利亞女王登基六十週年紀念，皇仁書院於年初接受大批少年新生，祖父廷羡公亦於當時入學，編號六七〇七(73)。

中央書院創辦宗旨為成為東、西雙方橋樑及兼顧到中國傳統文化。初期只收中國學生，且不硬性規定必修英文，但仍有許多不同種族、宗教的少年慕名而來就讀該校。儘管各族裔的學生穿不同的裝束－中國學生穿長衫、留辮；印度學生包頭；歐洲學生穿西裝打領帶，大家仍能共同相處學習。一八八〇年時學校已有十二國不同國籍學生，各國學生仍可保存本身的宗教信仰及習俗，例如回教學生在齋戒月可請假停課，中國學生在清明節，猶太人在秋節，基督徒在復活節均可回家過節[74]。校內經常可見中國學生翻譯依索寓言，其他國籍學生背誦三字經。皇仁書院校刊「黄龍報」每月均刊登孔夫子名言：「四海之內皆兄弟也」，令學生了解彼此理念及人生觀[75]。

祖父就學時期亦是實習教師。由於當年尚未有大學或師範學院，所以任何有意從事教學者，都須先行見習做實習教師。每日清晨須先上兩小時英文及學校教授法即教育學，其餘時間便必須教導約六十人一班的低班學生，課程包括中、英文基礎教育的讀、寫及算術；另外在學生家長許可情況下研習聖經，但僅為增進中英文翻譯技巧之用。

實習教師有時要與其他兩班學生共用禮堂一起教書，既無屏風隔離，亦無講台可站，在如此環境下應付六十多名學生實非易事。晚間則為自修時間，學校不供住宿，每月僅給十元或十五元津貼，根本不夠搭電車或坐轎，唯有步行往返。

在求學期間，祖父寄住於禮昌隆公司樓上（曾祖父與曾叔祖文奕公所創之公司），位於皇后大道中，離學校不遠。學校放假時祖父才回廣州鄉下嘉寮村與家人團聚。

日後部份實習教師繼續留校成為皇仁書院華人教師，也有人在國內擔任要職，亦有人如同祖父一樣，成為香港傑出菁英人士[76]。

一八九八年祖父十八歲已達適婚年齡，曾祖父母依當時中國習俗在鄉下請媒人相親。媒人會選擇門當戶對人家女子相配，一如：竹門對竹門，木門對木門，同時男女新人雙方生辰八字亦一

定要相配合。祖父在如此安排情形下，娶了台山鄉下女子黃蘭芳為妻。婚禮於新會嘉寮坊舉行，當時祖父仍在皇仁書院求學。

在皇仁書院的一九〇〇年頒獎典禮中，祖父曾榮獲特殊獎賞(77)。次年祖父離校跨入社會開始工作，但仍與母校保持密切聯繫。一九一〇年祖父與其他兩位校友共同對皇仁書院慷慨捐款，同年並設立利希慎獎學金(78)。

一九二五年祖父為紀念其好友江其輝先生，應允於皇仁書院另設獎學金(79)。

皇仁書院于一九二〇年再度成立業餘戲劇團，為戰亂飢民籌款，義演數場羅蜜歐與茱麗葉。祖父特頒授獎章給售票最多之學生，並出資支持皇仁樂隊在香港校際運動大會中演奏。由於他對母校各方貢獻，當年學校特將祖父肖像與Bertram Tanner校長肖像共同懸掛(80)。

一九二〇年十二月皇仁校友會成立，主要目的為協助母校成為校友及在校學生之橋樑，皇仁校刊「黃龍報」的編輯Kay先生，在第二屆校友會中特別在校刊做詩表揚祖父及其他校友會創始人；「…所有校友要對校友會創辦元老乾杯致謝，先祝Anderson及他的妹仔姨一路成功；還有我們的誠實大好人司庫Barretto；若是天明頭疼便是利希慎的錯…(81)。」祖父每年在校友會聚餐時均供應香檳酒，大家都稱為「利希慎歡樂汁(82)。」一九二八年祖父被推選為校友會主席。

利希慎公

依中國習俗，一人一生中常有兩個名字，一為出生時父母所命，年長踏入社會或成婚時，自己再取一名為號。祖父開始經商時為自己取名希慎，希望自己做人立業謹言慎行。廷羡此名除其父母及族譜記載之外，未曾再用。

祖父生前身材高大，神采奕奕，相貌堂堂，有典型中國面容，微黑皮膚，丹鳳眼，高顴骨，雖然他思想與做生意手法極為西化，而且英語流利，但平日衣着多為中裝，且隨中國習俗。祖父為人精明能幹，工作勤奮，富進取心，頗得人緣，又極樂善好施。

祖父第一個工作是由一九〇一年在陳念典先生經營的兆榮昌公司開始。該公司位於文咸街，專由中國進口中國藥材及乾貨，再轉銷至東南亞。由於祖父做事幹練，廿一歲時即為公司重用，擢升為經理。

祖父後來加入新加坡貿易公司所屬的雙德豐船務公司工作，處理香港與馬來亞之間的業務，並經常乘坐公司船隻往返各地。祖父總是密切注意生意機會，後來加入運輸行業，購下南亨船務公司，旗下貨船開始密集往來於香港、新加坡、馬來亞及仰光之間。在短短幾年時間，祖父已成為專營中國與東南亞轉口貿易的南北行行商中受人尊重的知名人物。

早期華商的刻苦工作、勇於創業的精神令祖父獲益頗大。由一八五〇年代開始有不同社團成立，尤以南北行公所的創立，令華商能彼此互助改進經營方式，並維持中國街坊治安(83)。

十九世紀初期，香港華人與歐人因種族隔離完全分開。一八六四年並訂下規章，公園內不許華人，轎子及未拴皮帶的狗入內，亦不許華人與歐人同時進入圖書館及市立博物館(84)。警察僅保護歐人及其產業，華人便自組社團僱用守衛以保護住家及商店。較為富有的社員因社區經常發生火警，便捐款購買消防救火車(85)。

一八七七年Hennessy成為香港總督之後，了解到要促進香港繁榮，必須有賴華人[86]。至一八八〇年時，華人在香港經濟發展中已舉足輕重。華人不只是香港最大地主，亦是香港殖民地政府百分之九十的稅收來源[87]，華商更漸成為歐商主要競爭對手。

在Hennessy成為港督之前，香港中區僅准歐人居住。曾祖父良奕公及曾叔祖文奕公於一八九六年携子由舊金山返鄉後，曾叔祖返開平鄉下定居，曾祖父則將家人遷至新會。日後兄弟二人赴港，曾祖父將祖父送至皇仁書院就讀。兄弟二人於皇后大道中二〇二號成立禮昌隆公司，曾祖父與曾叔祖經常往返香港與家鄉之間。祖父踏入商場之後，對禮昌隆公司投入大量資金以取得控制權，並購得其產業[88]。

禮昌隆由上海進口藍布銷售給香港華人，藍布即陰丹士林布，普遍為一般人穿着。祖父用大廈二樓做寫字間，三樓則供曾祖父及家鄉親友來訪時使用。祖父堂弟利樹源幼年在皇仁書院求學時亦住三樓[89]。公司業務一直經營至二次世界大戰末期方止，而此幢樓宇產業也一直屬利氏家族所有，直至一九九二年才轉手。

祖父平日多在早上去各銀行探訪朋友，如滙豐銀行及法國東方滙理銀行的皇仁校友何穎泉，另有在雪廠街花旗銀行的姚鉅賢等，這些銀行均多方資助祖父事業。雖然祖父的英文程度大可與銀行的西方經理直接來往，但他寧可交由自己朋友來做。

銀行訪友後，祖父才去禮昌隆查閱來往信件及處理業務。中午十二點半至一點之間去威靈頓街一九六至一九八號的裕記會所午餐，與朋友商討生意，下午再去證券交易所與他經常往來的Benjamin and Pots經紀公司與施遇文及高寶森等經紀人商討股市行情—祖父在許多上市公司投資甚鉅，如中華電力公司、中國糖房、牛奶公司、香港電燈公司、滙豐銀行、電車公司、均益倉及青洲英泥公司等[90]，到晚間則再返裕記會所晚膳。由於當時許多生意皆在會所商議成交，所以祖父

平日極少回家用餐，而且多半至夜晚十點才返住處(91)。

一八五八年鴉片買賣在香港合法化，祖父亦有投資於此。每年政府公報均根據一八五八年法令「殖民地鴉片加工及銷售特權」的條例公開招標買賣鴉片(92)。一九一二年四月廣東四邑人馬氏家族以每股一百元，兩千股共二十萬資本起家組成裕與公司。主要股東有馬持隆、馬敍朝。祖父與其他數人亦擁有裕與公司股份。該公司主要業務為經營鴉片。一九一四年其中一名股東對公司運作不滿，向法院提出控訴，要求法院下令公司停業。法院派員清查接收，但在其他股東極力爭辯反駁下，最高法院院長Sir William Rees Davies審終宣判裕與公司勝訴(93)。

祖父于一九一三年被推選為東華醫院總理。該醫院在香港政府協助，及海內外富有華人共策之下，於一八七二年成立，專為病重窮困華人服務。一八九四年大批華人死於瘟疫傳染，香港政府責怪醫院管理不善，故醫院除原有的傳統中醫外，也開始聘用華人西醫。香港政府於一九〇三年起接管醫院，並每年撥款六千元作為經費(94)。醫院設董事會，每年選出華人董事管理該院業務，各董事皆選自不同行業。後來東華醫院董事會漸成為香港殖民地華人權勢地位表徵，並與社區街坊團體及商會保持密切關係。醫院不但提供醫療服務，並兼辦社團慈善事業，殖民地政府當局首次有可靠機構，可聽取華人意見(95)。

一九二三年祖父取代Dr. J. W. Noble成為中華電力公司四人顧問委員會成員之一(96)。一八九八年Shewan Tomes公司買辦馮華川與朋友數人合組廣州市電力及滅火公司，經廣東省長批准，可對廣州及鄰近地區供電。後因公司財務困難求助於Shewan Tomes公司，導致中華電力聯合企業的成立。該公司於一九〇〇年由Robert Shewan以二十萬資本註冊(97)，分成二百股，每股千元。一九〇一年一月監管此聯合企業之顧問委員會，決定因現有資金不足，應另組公司接管業務，於是在一九〇一年一月二十五日中華電力公司改組為股份有限公司。祖父為公司主要股東，一九二八年時

已擁有43,416股[98]。廣州計劃流產後，此公司於一九〇九年售予另一中國公司[99]，中華電力公司至今仍為香港最主要電力公司之一。

祖父對香港年輕人極為關心，因而致力促成政府捐出嘉路連山地興建體育場，後被推選為首屆南華體育會會長。

專為教育香港女學生而設立的聖保羅女校，於一九二五年六月六日由港督Reginald Edward Stubbs破土奠基[100]。除由香港政府慨贈五萬元之外，祖父所捐贈三萬元為各人之冠[101]，是以學校將禮堂命名為利希慎堂。二次世界大戰之後，該校改為男女同校。我於三年班開始就讀該校有年，一直對母校保有許多美好回憶，當年幾位同學至今仍為好友。聖保羅男女校素以高水準成績及嚴謹校風紀律聞名，今日仍是香港最優良學府之一。

銅鑼灣

雖然祖父在事業上興趣甚為廣泛，但對房地產及地產發展則最為熱衷，並對香港發展目光獨到，他在中區及銅鑼灣購入大批土地及樓宇用來做大規模發展。

利希慎置業有限公司於一九二三年十一月二十七日，在高露雲律師行C.E.H. Beavis大律師公証下，正式註冊登記，設址皇后大道中二〇二號，電話二一四五，電報：香港利希慎。直到今日，此律師事務所仍為我們家族公司正式法律代表。公司營業範圍包括所有在香港的土地、物業、樓宇之買賣、租賃；對各種道路、鐵路、住宅、酒店、俱樂部、餐廳、公園、商店及娛樂場所之開發興建維修及管理經營[102]。

祖父在做任何重大投資之前，如購入銅鑼灣產業，均先深入研究。十九世紀時銅鑼灣是香港這個荒島上最美的一塊高地，四周除海港外，山陵環繞，由海港可遠眺大陸。附近的鯉魚門深水海港，入口狹窄受環山陡壁保護，可阻擋強勁西風，供大船停泊。

一八四一年遠東首家英國商行Jardine Matheson and Co.即怡和洋行在銅鑼灣開始營業。Dr. William Jardine、James Matheson及Hollingworth Magniac三人為生意合夥人。該公司於一八二〇年間即在澳門展開業務，並有意將業務拓展至印度。Dr. Jardine是來自Dumfriesshire的蘇格蘭人，在東印度公司服務時曾行醫。Matheson為Donald Matheson指揮官的次子，世代定居於Shinness Sutherland[103]。Magniac為富有法國大家族de Magnac後裔，信奉法國新教，在一六八五年時因受法國政府迫害，便由法國逃離至英國[104]。約在一八〇五年左右H. Magniac赴澳門，與當地經營已久的Beale, Read and Co.來往。他當時或許也是Charles Magniac公司的合夥人。一八二四年Charles死後，公司改為Magniac and Co.由Hollingworth及Daniel兄弟二人共同經營[105]。Magniac在東印度公司旗下持有執照為怡和洋行的澳門代理商[106]。早期他們眾人所建立之商業

網為：Jardine負責打通中印通商管道；Matheson駐留印度銷售遠東貨品；H. Magniac則為廣州、澳門銷售代理。至一八二七年因業務發展迅速，Jardine及Matheson決定須定居澳門就近照顧業務，每逢交易季節再北上廣州。早年一直由「掛牌」代理行Magniac公司代理經銷業務。Jardine及Matheson分別於一八二六年及一八二八年加入公司合作，至一八三二年六月三十日Magniac公司解散後，怡和洋行於一八三二年七月一日成立[107]。

Charles Elliot指揮官為實現將香港成為廣州第二的理想，首次於一八四一年六月十四日公開拍賣五十批土地以發展香港。由於當時行事草率，對日後產權造成極大困擾。當怡和洋行由英國政府買下銅鑼灣產業時，原本以為可擁有土地所有權，或至少有長期租約，豈知當時簽訂之買賣條約中僅有七十五年租約，在Alexander Matheson強烈反對情況下，終於爭取到改為九百九十九年期[108]。

大批的辦公樓宇、倉庫、公私住宅迅速在銅鑼灣興建，同時船塢亦建造完成，供各種商船入港停泊及整修，岸邊倉庫貨棧毗鄰，泊滿各類不同船隻。在海拔二百呎、風景優美、可俯覽海景的半山區，富商所擁有的豪華大宅開始出現。這些深院大廈佔地廣闊、林木扶疏、外牆多為石材。沿海貨棧設有炮台防守，抵禦流寇侵犯。

怡和洋行在一八六四年遷移至市中心後，其位於銅鑼灣的樓宇，改供一般職員使用。R. Jardine爵士於一九〇五年去世後，公司改組為私人有限公司。

大約在一九二三年間，祖父決定購買銅鑼灣產業時，洞察到汲水門港水淺，所有進入香港之大船均須經鯉魚門港口入港。又由英國港務專家在一九二〇年來港勘查結果指出，西環發展已達飽和，將來的發展勢必東移至九龍灣。祖父遂計劃由銅鑼灣供應建材，在北角填海以新生地建造房屋。由當時灣仔地價隨人口急增而大幅升值來判斷，北角遠景良佳。在北角沿岸如填海闢地，可由鄰近銅鑼灣山供應建材，極為方便，一旦計劃完成，便可立刻提供大幅新生土地作各項建築

發展。祖父經過詳細考慮後，認為此計劃有利可圖，預計可以一百○四萬三百六十九元成本費，填造出六十三萬六千○四十平方呎的土地面積[109]。

根據祖父的估算，新填海區可提供土地共建四百六十七幢華人住宅。更由於大批房屋同時興建，成本可低於其他香港營造商價錢，鋼筋水泥可自行進口，木材石磚可大量採購以降低成本。每幢樓宇的造價約為一萬一千四百元，每一公寓單位可收二十元月租，全部出租後可回收投資金八分利息。由於香港殖民地從未有此類租金低廉、且具有現代化衛生設備之堅固住宅，故對中產階層市民極具吸引力[110]。

一九二四年一月十七日，蘇格蘭Dumfriesshire, Comlongon Castle的鄉紳John William Buchanan Jardine將銅鑼灣土地以$3,850,960.35價格售予利希慎置業有限公司[111]。在此同時另一香港知名富商買進西環土地，並公開認為祖父對銅鑼灣投資判斷錯誤，但歷史証明祖父目光獨到[112]。購置銅鑼灣產業是祖父事業高峰之一，不只為家族購置產業、商業樓宇及由怡和大班手下買進貨倉，更代表著香港華人在逆境中所達到的成就。

當祖父購入銅鑼灣地產時，與香港政府協訂用銅鑼灣山泥土為北角填海，將來祖父有權發展新填的土地。但成交後，香港政府毀約，決定改用政府公地摩理臣山泥土，使政府有新生地產權，事後政府方知摩理臣山土質多為岩石，炸碎岩石極為費錢費力[113]。

祖父此時將銅鑼灣土地闢為花園及遊樂場營業，即利園山，並將大班留下的巨宅改為酒樓，整年開放，供華人遊樂，業務斐然。

利希慎置業有限公司於一九二五年將公司大部股權分配家人名下：祖父一千股，兒子每人五百股，祖母二百五十股，三位妾室各一百股，女兒各七十股，另將一至五的散股配予其他親戚[114]。

祖父對置業公司組成的確非常明智。直到今日，子子孫孫都十分佩服他的遠見，因家族成員

均為公司股東，須大家團結合作共謀利益，日後家族後裔將配股規則修正以求更為公平。

祖父後來買下銅鑼灣山下原為貧民區的土地，將街道拓寬，重建屋宇。一九二六年一座金壁輝煌而具有當時先進旋轉舞台的中國劇院－利舞台，巍然聳立於波斯富街上，為當年廣受歡迎的娛樂－中國戲劇提供了演出場地。劇院的拱形圓頂繪有金龍，掛滿燈飾，確是當時香港最豪華之中國劇院。並舉行了一場對聯比賽，選出皇仁書院中文教師李精一先生所撰寫之對聯，懸掛於舞台兩側(115)。

利舞台廣場於一九九四年在利舞台劇院舊址重建完工，新建大廈仍保留利舞台劇院原有風貌，同樣有高聳拱形圓頂。李精一先生所寫的對聯在施工期間特別移開妥為保存，現在則高懸於利舞台廣場新廈大廳正中。

祖父日後在自己本人、祖母及置業公司名下大量置產。除銅鑼灣、堅尼地道自宅，及皇后大道中的自用寫字樓外，祖父其他產業亦遍佈於波斯富街、利通街、灣仔道、太和街、石水渠街、皇后大道東、海旁東、第二街及春園街等(116)。

二次世界大戰後銅鑼灣山地逐漸鏟平建屋，山石泥土則用於填海。原本銅鑼灣範圍是東至渣甸街，南至禮頓道，西至花園路，北至軒尼詩道。今天希慎興業有限公司在該區及其他地區擁有大批地產物業，包括：嘉蘭中心、友邦中心、禮頓道111號、興利中心、新寧大廈、新寧閣、禮頓中心、希慎道壹號、利舞台大廈及重建之利園酒店等。

家族

祖母利黃蘭芳是家中女家長，生於一八八〇年六月三日，比祖父年長三個月，我們叫她「人人」。她身材高大，面形飽滿，婚後數年未能生育，後生一女，又不幸夭折。

纏小腳的祖母娘家相當富裕，在當時，社會風氣認為纏足女子代表此女家境良好，毋需工作。她們由嬰兒期就開始將腳骨綑綁，阻止發育，硬生生的將腳掌骨骼纏扭至僅有四吋左右的所謂三吋金蓮，這些女孩為著家人身份地位須強忍痛苦。以前中國男人覺得小腳性感，又愛看如同風吹柳擺的小腳女人走路，男人認為如此方為搖曳生姿別具媚力。我一直記得祖母舉步艱難的模樣，雖然後來解放纏腳，但祖母雙腳已是扭曲不堪完全變形了。

祖父雖然身受西方教育，但許多方面仍保有中國傳統，不像當時許多其他男人遺棄父母安排的元配，他仍尊重祖母在家中超然的地位。由古老相片中可看出，除非是家族團體照，祖父多半只與祖母，或與祖母及子女合照，並將數筆產業置於祖母名下。祖父有吸食鴉片習慣，當他知道祖母對鴉片過敏之後，從此不在家中吸鴉片。

雖然祖母未曾讀書，一字不識，但祖父仍請教師花了很長一段時間教祖母用英文簽名，祖母稱寫英文字是畫雞腸。二次大戰後，我們曾見過這位東印度女英文教師，她常在舊曆年時來向祖母拜年。

祖母為人非常儉樸，亦從無笑臉，也不大聲說話。二次大戰後所有家人同住一處，我們常去看望她，她總是坐在同一張椅子，面露愁容悶悶不樂。

祖父後來在廣州市納妾，即是姨太或二太。當妾室進門之時須對正室大房叩頭奉茶示敬，此後才能成為家庭一份子。二姨太身材纖細，面孔尖瘦，和一般的妾室一樣，家境清寒，須要工作，所以從未纏腳。二姨太在城市長大，比祖母多見世面，見聞較廣[117]，除此之外似乎無人知道

她的家庭背景。以前妾室入門之後不可回家探訪親人，亦不許有親朋來訪，她們是家中財產的一部份。二姨太十分認命，從不反抗，從不爭辯，甚至連她親生女兒對她背景都毫無所知，只知二姨太比祖父年幼七歲，她是當時舊制度下的生存者，日後贏得所有家人的尊重。

當二姨太傳出懷孕的消息，家人均極為興奮，祖父更特別請洋人助產士接生(118)。這是祖父婚後七年首次傳出喜訊，我想祖父不願在生產過程中有任何差錯發生，又認為歐洲助產士較中國接生婆經驗豐富及衛生，故此不惜聘用洋人助產士。一九〇五年三月七日，二姨太生下長子銘澤，我的父親。

父親出世後福星高照，不只是家中有子繼承香燈，祖父事業亦一帆風順。父親承由祖母親自撫養，祈能因此借福生男育女，果不其然，祖母也陸續懷孕生了二子、二女：排行三子榮根（又名孝和）、四子榮森（J. S.）、二女舜華及五女舜賢。祖母與父親感情融洽一如親生母子、祖母對父親任何決定均十分尊重，日後對家族裨益極大。

二姨太後來陸續生下次子銘洽，及三個女兒，她們為三女舜英、四女舜琹及六女舜豪。

祖母在家中掌控一切家務，名義上是所有子女的母親。妾室所生之子女均須尊稱正室為「媽媽」，生母為「阿姐」。依中國律法，妾室所生子女皆為合法子女，享有平等地位。妾室們有時經由主人准許可離去，或在主人死後，正室要姨太們離開，但絕不可將他們的子女帶走，孩子屬於家族的成員。

二姨太對祖母非常尊敬，從不與她爭論，常以天知地知的藉口避開無謂之糾紛(119)。記憶中二姨太總是精力充沛，常說我們若是不乖，她會活剝我們的皮，其實我們知道她做夢也不會做這種事。她是個好母親，好祖母，在世時候不僅對子女，甚至對孫輩都盡心盡力(120)。

雖然二姨太有傭人侍奉，但在過年期間她喜歡在院中用傳統老方法蒸炊年糕。我見過她蹲在

大灶前用吹火管生火，總覺她仍然住在中國鄉下。糕蒸好之後分派每個子女家中，父親每年農曆生日的時候，她還會為父親做芋頭鴨，她說是父親自小最愛吃的家鄉菜。

我以前最愛看她抽水烟，她將烟絲塞入精緻盛水的景泰藍煙筒管口，然後用根點燃的紙捻去燒烟絲。當她吸的時候，煙氣就經水廻旋吸入，發出「咕嚕」聲響，她然後吐出餘煙。我們這些孩子們聽到水煙筒咕嚕咕嚕，就哈哈大笑，是我們最喜歡的娛樂節目之一。

二姨太每天在佛堂拜佛，她在跪拜神台之前必先換上咖啡色粗布袍。神台不只供奉菩薩，還有其他神像，在燒香之餘亦須獻上鮮菓及當晚她所要吃的食物。她唸經時手捏佛珠，聲調聽來非常單調乏味。我們幼年時常在佛堂嬉戲玩耍，當二姨太唸經時常在她背後扮鬼臉，若給她知道的話非罰不可，但在她唸經的時候決不會被發現。

二姨太晚年時將所有祖父買給她的首飾轉贈給女兒，她認為沒有必要再穿戴首飾，若有特別場合需要的話，她再向女兒借回(121)。在她七十歲生日之後，她將所有存款（由家用省下的私房錢）分給兒孫。記得多年前我曾分到七百元，她在過世多年前即將壽衣及後事安排妥當。

祖母於一九五六年過世後，二姨太便成為家中的女家長，八十歲以後還首次出國遠赴舊金山，雖然她從未坐過飛機又不識英文，但她認為有菩薩保祐，一點也不害怕。二姨太生前健康，知足，一輩子抽水烟，近百歲時死於肺癌。

當祖父在香港財富日增，聲望日隆的時候，他聽到傳言謂嘉寮父母受土匪威脅綁架勒索，便將父母遷至新會，在竹林里二十八號築屋奉養(122)。

新屋為當時典型大戶人家宅第，一進大門即是畫滿壁畫的天井，天井右方門內設置神台香案

供奉各方神靈（神台在文化大革命時被摧毀）。天井四周排水管頂飾有彩陶金魚，雨水可經魚尾由魚嘴泄至排水管。

大廳對門的雕花大理石供桌上列有祖宗牌位，桌上香爐旁有四時鮮菓。每逢年節時則用不同食物供祭祖先。大廳兩旁列古董太師椅，其中一張最為特別，因為據傳坐過這張太師椅的人，必有能力將家族事業領導發展。我們在一九八〇年初回新會時，父親要漢釗堂兄（銘洽二叔長子，祖父第一個孫子）坐在這張椅上，漢釗堂兄日後証明此說果然不虛，而顯然父親以前也曾坐過[123]。

大廳兩側為臥室，屋後有廚房及儲藏室，後院有口水井，廁所及浴室與主宅分開另建，房舍側邊築有高牆圍繞的花園，天井有階梯直通屋頂可俯視全村景色。

這幢祖傳老屋雖經共產黨戰亂，及文化大革命摧殘受損，但外表仍保持原貌，一直由新會親戚照顧，只是不再有人居住。

曾祖父母在世住此屋時，常有親戚到訪。父親幼年時候，曾祖父常常對他敘述他在舊金山的經歷，及太平洋海上艱辛的旅程。由於國人「民以食為天」的觀念，曾祖母甚至在父親熟睡的時候都要塞雞腿給他吃，以表示對父親的寵愛[124]。

當曾祖父過世的時候，依鄉下風俗，須將遺體由嘉寮村沿河坐船出殯，由父親坐在船頭，祖父坐在船尾，曾祖父的棺木則放在船中間[125]。他們所選的墓地甚難送達，須經一日跋涉之後再爬上山頂[126]。據說墳地風水甚佳，會蔭澤後代子孫。

一九一〇年時香港流行鼠疫，衛生情況頗差，祖父便將家人遷至澳門，住荷蘭院正街至一九一八年[127]，（此區於一六二〇年荷葡戰役時為荷蘭人之戰俘營，因而得名）。祖父對子女教育問題極為重視，故在他們離開香港往澳門期間，特聘請陳子褒先生回家為子女授課[128]。

父親返回香港之後就讀於皇仁書院，十五歲便與幼弟榮根（又名孝和）遠赴英國求學，住英

藉監護人Mr. Churchill家中，由私人教師教授課業。後來兩位妹妹舜華、舜英也相繼赴英，入女子住宿學校就讀。舜華是祖母所生，舜英為二姨太女兒，兩人年齡相差衹有九個月。在赴英國時，由於旅行文件引起麻煩，為避多費唇舌解說中國多妻制度風俗，便將出生日期改變。

求學期間父親英文名字為Richard Charles（朋友稱Dick）、榮根為Harold、舜華為Doris、舜英為Ansie。父親後來入牛津大學研讀土木工程，榮根叔父則專攻法律。

第一次世界大戰期間，祖父在灣仔之山坡地購下一大片土地，為自己及家人建造住宅，但因戰爭及勞工問題拖延至一九二〇年才開始動工。這幢由Palmer and Turner公司負責設計建造，位於堅尼地道三十二號（後改為七十四號）的大屋，當時電話為二四五五。十九世紀時，極少中國家庭可擁有一幢如此氣派的住屋，在祖父事業生涯來說是另一高峰。

位於堅尼地道的這幢產業四周築有石牆圍繞，內有花園、果樹、菜畦、雞舍、寶塔、噴泉、金魚池及假山岩洞，後來又加建網球場。由花園可遠眺香港，景觀極美。庭園由多位專人照顧，整年花開不斷。

大屋門口全日有印度籍配槍武裝警衛看守。當時香港人認為印度包頭錫克族人相貌兇狠，鬚髮較中國人濃密，多僱用為門房守衛。家中的守衛及其家人，均分住於花園觀音像噴水池邊的房舍，世代跟隨我們工作。

二次世界大戰後一個叫南星(Nam Singh)的守衛與我最熟。那時我不知所有印度錫克族人都姓星(Singh)，他們廣東話都很流利，平日除了持槍故做兇狠狀外，以乎別無他事。南星都值日班，我們在花園玩耍時，常與我們談天。記得有一次他給我看他在花盆下捉到的蛇，還說他預備換班之後煮來吃。

我們家住堅尼地道大屋多年，我亦出生於此，對大屋有許多美好回憶，當時堂兄妹亦同住一

起，自小不乏玩伴，每天可在花園不同地方玩耍探險，一年到頭都可以在院裡各種不同的中國南方菓樹上採摘新鮮的水菓來吃。當我們想吃木瓜的時候，有人就會鋪開一件雨衣，然後把著樹幹將熟木瓜搖下來。若搖不下來，就用棍棒挑下來，用雨衣接著，大家就毫不客氣的大吃裂開的木瓜，吃完之後一定在院中洗乾淨嘴臉後才敢進屋，以免受到責罵。

春天時我們用手帕在金魚池中撈蝌蚪、雞棚趕小雞。夏天就捉知了、撲蜻蜓，有一種樹上的昆蟲如蜻蜓大小，有長鼻子，我們捉來之後用線綁著它鼻子飛，當紙鷂玩。整年在花園爬樹、吊樹桿、採水菓，當自己是人猿泰山。又在園中蕩鞦韆、溜滑梯、踩高蹺、騎著單車到處亂穿梭，每天傍晚當佣人四出找我們回去沖涼的時候，我最喜歡躲在假山洞裡。

「大屋」是當時香港最雄偉的住屋建築之一，前廳可容納至少五百多位賓客，一樓有大型宴客用正式餐廳，幾處客廳及大廚房為宴客使用，後面為工人房，二樓及三樓為家人使用之起居間，都有大洋台，還有足夠的客房供鄉下來訪的親友使用。

周末時祖父多半在家與家人及鄉下來訪的親友同枱用飯，婦人同年幼子女在三樓，男人、青少年則在二樓用飯，由廚房同時配饍(129)。

大屋內擺滿世界各國藝術品，但很多收藏品在二次世界大戰家人棄屋逃難時被竊，所幸其中一件極為珍貴之整支長毛巨象象牙雕刻得以倖存。此件置於雕刻精美的黑檀木座上的藝術品，原屬一名廣東法國商人所有。他購入整支象牙材料後，請名雕刻家梁維，將三國演義故事雕刻其上，經多年工夫梁維始將唯妙唯肖的三國人物雕刻完成。後來這個法國商人因個人財務困難，而將此藝術品抵押給法國東方滙理銀行貸款周轉。透過該行買辦何穎泉的關係，祖父於一九二六年

購得此件藝術品。我幼年時在大屋前廳玩耍時常將這支大象牙當馬騎，而從未有大人阻止，回想起來都覺得驚訝。這件精美的藝術品後來捐給香港中文大學博物館陳列。

大屋樓下有幾張用黑檀木細雕的煙榻床，我們這些孫輩的孩子們常爬上榻床，大家擠在一張床上講故事。

大屋正廳非常光亮，但上二樓及三樓的樓梯邊間，因木板深色，總是昏昏暗暗。二次大戰後，孩子們一起玩的時候，大孩子常追上樓梯扮鬼嚇唬我們，孩子當中以我最小，爬樓梯也最慢。

大屋完工之後，祖父在鄰近山上又興建另一大廈－利行。大廈外表雄偉，中間有開濶階梯，內分六間明亮寬暢單位，專出租給日本、歐洲租客。雖然大屋與公寓大廈之間用幾個花園相連，但我們很少見到租客出現(130)。二次大戰之後，利氏家庭成員漸多，利行亦收回由家人使用，不知祖父當年是否想到將來他的子孫也會住進這幢大廈。我們搬至利行後，父親與母親仍常在大屋中招待生意上的朋友，孩子們依舊在大屋花園玩耍。

大屋落成時祖父已有兩房妾室。當年中國男人均以子女及妾室人數來炫耀財富。

三姨太是祖父在常去的茶樓看上娶回的戲子(131)，她嗓音很尖，圓圓臉孔，笑起來很甜。

祖父母、姨太們及所有子女都同住在大屋，祖母及其子女住二樓，二姨太、三姨太及其子女住三樓。因三姨太常與祖母爭執，便遷出到銅鑼灣波斯富街公寓，和她兒子榮杰（排行五子）同住(132)。她甚至在祖父過世之後都沒有再搬回大屋，我一兩年才見到她一次，所以對她認識最少。

一九二〇年初全家搬入大屋之後，祖父又娶進四姨太。她是祖父幾個有錢朋友在聲色場中宴請祖父時遇見的，當晚鴇母聲稱手下有一新到女子仍屬姑娘，可陪侍化鈔最高的紳襟，結果祖父以最高價贏得此女。日後祖父顯然對她難以忘情，便將她娶回做妾室(133)。

四姨太從未住過大屋，她在祖父寫字樓附近擁有自己住處，祖父常在白天去看她，有時四子榮森也會同行。四叔仍記得他不舒服的時候，四姨太對他十分照顧。後來四叔留香港讀中文學校，子女中唯有榮森叔父常隨祖父查察生意，造訪朋友及往來客戶(134)。

四姨太身材瘦長，母親與父親成婚之後，與母親成為好友。常對年幼十歲的母親傾訴心事，覺得母親是家中唯一與她沒有利害關係的人。她對母親透露，去歡場賺錢是她自願而非家庭所迫，認為窮家女唯有在歡場才有機會遇見富人，嫁入豪門為妾，擺脫貧窮。她害怕窮困，不能想像自己為窮人妻的窘境。多年前我曾去她家中探望她，當時她已九十多歲，但仍很健朗。談話中我提到她孫女在多倫多近況很好，而且頗為快樂，她問道：「她是不是賺很多錢？」這時我才了解錢對她的重要性，有錢她才有安全感。

祖父過世時四姨太仍非常年輕，有離家再嫁的自由，她告訴母親她選擇留家守寡的理由是她難捨她的子女。後來她長期吃齋拜佛，與許多尼姑成為好友。二次大戰後她搬入利行與她的子女，六子榮康、七子榮達、七女舜儀、八女舜娥同住。

祖父的幼堂弟利樹源九叔公是文奕公的兒子，對祖父十分懷念。當樹源叔公在皇仁書院求學住皇后大道中二〇二號時，星期假日必來大屋探訪祖父母，並留下與其他親人一起吃中飯。他說祖父為人嚴肅，對下代年輕人要求循規蹈矩，勤奮努力，大家對他都十分敬畏。

利舞台完工後，祖父常在飯後帶樹源叔公去利舞台看戲。他們談到未來戲院理想時，常說有朝一日一定可以躺著看戲。樹源九叔公認為此即當今之在家看電視之預言。

在樹源叔公心目中，祖父一向以勤奮工作及努力求知向學為做人目標。對不合他要求的人十

分嚴肅，但對自己親屬非常仁慈。他不但對父母盡責，對其他有需要的近親遠戚子女們以及同鄉，他都會妥善幫助照顧。祖父並供外甥、姪女讀書受教育，為兄弟安排工作，弟婦們終身均領有安家費，甚至祖父死後，祖母仍照祖父生前意願供給。

祖父對他所有親人均提供免費住處，當我在一九五九年初來加拿大時，遇到一位多年前隨家人由中國南方移民至渥太華的老太太，經由介紹認識後，她告訴我她有個親戚嫁給香港利氏家族，一直住在祖父提供的免費公寓裡。人總是會記得這種非比尋常的恩惠。

家人一直在大屋住到一九四一年日本佔領香港之後才搬至中國大陸，戰後返回香港時，祖母與其子女仍住大屋。除三姨太與其子榮杰住銅鑼灣之外，其餘家人則住利行。一九五〇年利園山遭鏟平發展，開拓興建新道路及多幢高樓大廈。後來我們一家搬入希慎道的使館大廈頂樓居住，這幢大廈是我們家族中所擁有的第一幢高樓大廈。在加路連大廈與崇明大廈相繼完工之後，其他親人亦陸續搬入。至於位於堅尼地道上的大屋與利行，在一九八五年由希慎興業有限公司改建為竹林苑綜合大住宅區，至今仍有多位利氏家族成員住於該區。

雙親—銘澤與瑤璧

祖父在一九二〇年時已將兩子、兩女送到英國求學，臨行前向兒女告誡，在國外不得與外籍人士論及婚嫁，否則將喪失遺產繼承權。

一九二三年父親負笈牛津大學研讀土木工程，當時英國留學生在英國均受良好照顧，不像在香港反遭外人歧視。在求學期間父親結交了數位日後成為世交的好友，建立珍貴友誼，多位同窗好友在回國後，均有卓越成就。

一九二五至一九二六年間，父親是歐洲中國留學生總會會長，同學之一錢昌照先生在大學讀經濟，回國後出任北京首席經濟顧問、教育部次長及全國政協副主席。父親一直珍藏當年他與同學們聚餐會的一張菜單，在這菜單之上，有他們共同草擬的新中國未來大計(135)。

一九二七年父親於牛津大學畢業後返香港探親，竟遭受到在英國所無的種族歧視，他覺得非常訝異。事緣於他留英期間經常騎馬，返港後亦想加入香港賽馬會繼續策騎，但卻因身為華人而被拒。祖父立刻預備自組華人賽馬會與之抗衡，香港賽馬會唯恐如此一來必會失去他們主要經濟來源－華人賭馬的豐厚利潤，便決定准許父親在俱樂部騎馬(136)。

父親原本亦想加入香港會，但也因身為華人被拒。這是他終生難忘的恥辱，多年後香港會邀請父親成為會員，但為父親所拒。

祖父一心希望身為長子的父親能在返英做實習工程師之前，完成終身大事，所以曾介紹許多年輕女子給父親認識，但他都不表興趣。連祖母也為父親相親，見了許多女孩。我曾聽祖母說過，當時有個女子她不喜歡，因為她腿毛太長！

一九二七年夏天，母親瑤璧與妹妹瑤蓮由九龍打網球回家途中，見姐姐瑤珍與一群年輕人在海邊野餐，便決定過去打個招呼。當時父親也在其中，便走向瑤珍要求她介紹認識這兩名剛加入

的少女，瑤珍說：「別煩她，她是我小妹。」但在父親堅持下，終於結織母親，兩人相處一下午後，由父親駕車送母親回家。

在雙方家長贊成下，兩人繼續交往數月，經常見面。父親喜歡邀約一群朋友一起玩，雖然他本人不愛跳舞，但經常請母親及其他朋友一起在香港酒店跳茶舞。

母親出身在一個非常新潮的中國家庭，我外祖父黃茂霖公英文流利，在香港頗受尊重，是一九一三年創辦的香港華人會計師公會首居主席。一九二三年外祖父被委為太平紳士，一九四一年六月十二日榮獲英王喬治六世授予榮譽勳章，又在民國初年由國父孫中山先生頒授四等嘉禾勳章，表彰其多方貢獻。外祖母麥玉珍是出名的歐亞混血美女，當年在香港的歐亞混血兒在生活習慣作風上，不是極端歐化，就是徹頭徹尾的中國化，而外祖母即是傳統中國式女人，熟知中國民情風俗，是人門經常請教的對象。

黃家全家都是基督徒，母親的祖父母極為慈祥。記得常聽母親提起他們兄弟姐妹及堂兄妹在曾外祖父菓園及別墅的種種童年趣事，該菓園及別墅位於當今的九龍太子道。曾外祖父曾在香港政府任法庭傳譯員公職(137)。兒孫輩最喜歡曾外祖母，並常提起她。她是個新時代女性，除中文之外，能說流利英語及德語。她鼓勵她的子女與洋人結婚，因為她要最漂亮的孫子。她喜歡唱歌，也愛聽孫兒唱給她聽，孫兒為她唱完讚美詩後，必得賞金。她對慈善公益一直非常熱心，甚至中風後仍坐著輪椅出力(138)。曾外祖母晚年時，累於不停生育，便為曾外祖父納一年輕妾室。曾外祖母共生十六位子女，外祖父是次子。

外祖父母共育子女十一位，母親自幼即進拔萃女子書院就讀，英文流利。黃家女子（母親姐妹及堂姐妹）在一九二〇年代是香港出名的美女，初見父親時才十七歲，仍在高中就讀，尚未涉足社交場合。

外祖父是香港第一個擁有汽車的人，母親與姐妹們常在香港到處開車，並告訴我們曾因開車超速被警察攔截。她以前喜歡打網球，也曾受飛行駕駛訓練，但從未考取飛行駕駛牌照。

父親後來對母親坦白表示他是庶出，非祖母親生，不知母親是否介意。由於父親與祖母感情融洽，一如親生母子，外人均以為父親為正室長子。母親表示並不介意，因為雖然自己家庭十分特殊，外祖父從未納妾，但亦明白多妻制當時在中國甚為普遍(139)。

祖父母對父親的選擇非常欣慰，在旋風式的求婚過程中，由於祖父急切盼望身為長子的父親，能在返英受訓前完成婚姻大事，所以當父親向母親求婚的同時，祖父母亦心急的同時拜訪外祖父母，徵求他們的同意。

唯一問題是年僅十七歲的母親，仍在高中就讀，希望畢業之後再結婚。她知道婚後將嫁入一個非常傳統的舊式家庭，所以不想太早成婚失卻自由。但憑著祖父能言善道的口才，終於說服母親，他保証婚後母親可與父親同赴英國牛津大學進修，並建議她學葡萄牙文，日後生意上可做他幫手。祖父並答應次年陪同她父母一起去歐洲探訪他們。事實上，他只要母親肯出嫁，對母親的所有要求一概允許，難怪母親對祖父一向如此尊重有好感！

父母的婚期訂在一九二八年二月廿八日，香港日報報導Swann牧師首次為華人新人利銘澤先生及黃瑤璧小姐，在聖約翰大教堂主持婚禮。

結婚當日教堂內外中西賀客雲集，坐無虛席。新娘、新郎乘坐全新飛雅特禮車由伴娘花僮陪同抵達教堂，新娘在銀紗鑲珠禮服及白玫瑰花襯托之下顯得格外動人。伴娘、女嬪相及花僮均穿綴滿小花的彩色絲綢禮服，配粉紅玫瑰花，雙方父母皆穿著正式中國大禮服。

慶祝酒會在利園舉行，中外賓客逾兩千，酒會特搭一個大竹棚（臨時由竹子及草蓆搭成），由香港酒店供應菜餚，竹棚內枱架上放置六層結婚大蛋糕。

婚禮中何東爵士代表賓客向新人敬酒致賀辭，他不但是祖父及外祖父好友，與曾外祖父黃有傳先生亦是世交摯友。在他精采演說中提到父親在一九二五至一九二六年為英國及愛爾蘭中國留學生總會會長，及他對中國的情感，孰不知因這愛國情操，使父親日後為中國前途做出各種貢獻。

原本預定由周壽臣爵士代表來賓向新人雙方父母敬酒致辭，但周爵士到會不久即因身體不適須提早離席，臨時由Dr. Robert H. Kotewall－羅旭和博士代替。

羅旭和博士亦為雙方家族多年好友，自是樂於替周爵士代表來賓致辭。他深知祖父及外祖父對香港貢獻良多，並稱新人雙方母親皆為典型賢妻良母，兩位新人出身此等家庭，相信必能同心協力，白頭偕老。

父親對何東爵士、羅博士及來賓致謝辭時表示；對大家給予的祝福與鼓勵非常感謝，更對所有親朋好友光臨婚禮感激不已，他與母親會將這大喜日子永誌在心。

婚後父親與母親乘船經蘇伊士運河赴瑞士渡蜜月，順道一遊埃及策騎駱駝、在獅身人面象前留影，非常愉快。到瑞士之後與孝和弟、舜英、舜華兩位妹妹會合，蜜月結束後同赴英國。

審判

一九二〇年代末期，祖父的其中一項事業捲入法律糾紛。在一九二七年適值長子剛由牛津大學畢業返國探親之際，裕成公司的突然結束實在是始料不及。

在一九二七年六月之前三年期間，裕成公司在澳門享有鴉片獨家專賣權，可在澳門加工並存有鴉片，同時可自由輸出輸入。十九世紀末至廿世紀初，香港政府及澳門葡萄牙政府均直接涉入鴉片貿易，政府除出售鴉片買賣牌照外，亦公開發牌給合法的煙館，政府與走私煙販競相殺價搶市場。

裕成公司於一九二四年成立，公司所有權為祖父兄長之長子利裕成一人所有（此公司與馬家數年前倒閉的裕興公司無關）。裕成公司成立資本三百萬，三分之一屬於利氏家族。祖父為該公司總經理，在三年間，裕成公司已擁有自己鴉片種殖場，三百萬的資金中已有四分之一回收。

根據一九二四至二五年日內瓦會議協定中的鴉片協議第六項條文規定，各簽約國同意其政府不得將鴉片種殖場租賃給任何人使用，並嚴禁將加工或未加工之鴉片銷往任何地區。

一九二七年三月廿六日，葡萄牙政府在Boletin公報上宣佈，根據一九二六年三月二日修訂通過，在一九二四年十一月二日至一九二五年二月十九日期間於日內瓦舉行的日內瓦鴉片會議協定：任何有關鴉片進口、加工及買賣的專利合約將會完全取消，因此澳門葡萄牙政府與裕成公司之間的合約將於一九二七年七月一日終止，改由政府專賣，並由消費稅督察監管。繼後由澳門法院命裕成公司結束，並接收鴉片種殖場，令裕成的投資化為烏有。

政府消費督察成立鴉片管理委員會，由羅保(Pedro Jose Lobo)出任行政管理官員。

一九二七年六月時祖父聽到謠傳，謂澳門政府將鴉片生意交給由祖父以前的手下－裕成公司經理呂儉畢及兩個惡名昭彰之走私販所組公司專賣。祖父後來又查到此公司名為又成公司，設址

於新成立的鴉片管理局的同一大樓內，並在香港成立辦事處，在香港、澳門及廣州招股吸資，並聲稱該公司已由澳門政府手中得到鴉片專賣權。

祖父是在陰錯陽差的情況下得知又成公司的消息的。原來又成公司在香港印度商業銀行亦設有戶頭，銀行信差弄錯買辦指示，將一張原送交澳門鴉片種殖場，而又蓋有又成公司簽章（已背書）的支票，誤送交了祖父。經祖父連日追查之下發現，又成公司在銀行開戶數日之後，即通知銀行又成公司已獲得澳門鴉片專賣權。

馮祥先生是祖父朋友，曾在裕成公司投資，在一九二七年八月，馮徵詢祖父對投資又成公司之意見。祖父建議他應先對該公司多做了解，馮祥便致函又成公司查詢資料，並要求書面回覆。

祖父在聽到澳門政府的各種行動之傳言後感到非常詫異，他認為澳門政府既終止裕成公司的所有合約，便不應再發牌予其他公司。

因此祖父立刻對澳門總督遞交陳情請願書，詳述他對澳門的各種服務及貢獻，並重申裕成公司一直遵照合約規定履行義務，但卻遭結束營業，祖父盼望政府能對其本人及裕成公司給予公平合理待遇。在請願書中，祖父亦提到澳門政府公佈之新鴉片政策，並稱澳門政府對裕成公司及其職員之行動，在葡萄牙法律之下應屬無效；祖父亦提到鴉片管理局未知會他，便即將鴉片專賣權授予呂儉畢；另外羅保涉嫌干預香港政府，為呂儉畢所受刑事控訴脫罪等事；祖父又指出澳門政府未在限期之內退還裕成公司所繳之鴉片合約保証金；及又成公司宣稱已取得澳門鴉片專賣權作吸取資金手段。基於以上事實及其他多項理由，祖父要求當局對此事件深入調查。

祖父將陳情書分別送交澳門總督、立法院、十位律師及葡萄牙駐香港領事Albuquerque。祖父又另請張才先生將馮祥所收到的回信譯為中文。

一九二八年春天，澳門的羅保控告祖父誹謗，並要求法庭下令禁止祖父繼續散發他對總督之

請願書，因請願書暗示又成公司以重金賄賂政府官員，以取得澳門鴉片獨家專賣權。

誹謗案主要涉及之澳門鴉片專賣權，及澳門又成公司以十二萬代價取得鴉片專利之控訴。原告認為這些誹謗會令他身為政府鴉片管理委員的身份失去威信。

最高法院之聽証會由首席法官Henry Gollan主持。原告由C. G Alabaster及G. H. Sheldon兩位律師代表，被告則由Eldon Potter及F. C. Jenkin代表出庭。

控方律師Alabaster質問祖父印製為數八百份之請願書，是否意圖散發至世界各地，祖父答稱即使印少於八百份，印刷費一樣也是四百元。

馮祥致祖父的信函亦被紀錄在案，該信指出：「根據又成公司對股東報告稱，已經由阿岳付出七萬元保証金及經羅保付出五萬元給財政部。可見又成公司是經不當手段，利用阿岳及鴉片管理委員羅保之關係，未經正式公開投標而取得鴉片專賣權，澳門政府縱容此公司為所欲為獨霸鴉片市場。」

一九二八年四月十七日高等法院首席法官Sir Henry Gollan宣判祖父無罪，判決中指出，陳情信對原告確有損害，但請願書中所舉例証均有事實根據，而且原告無法証明祖父將陳情信付印公開有惡意中傷他人意圖，故不能構成誹謗罪。官司結束後祖父頓鬆口氣，希望澳門總督能公平處理此事(140)。

在官司纏訟期間，祖父曾收到恐嚇信，威脅祖父停止散發陳情信，否則對祖父不利，並聲稱會於其大兒子婚禮中投擲炸彈。雖然朋友們均勸阻祖父多加小心，但祖父對這類恐嚇全置之不理。據樹源叔公說祖父從不擔心害怕，還嫌朋友過慮。若當年他稍加留意，必會將日常出入的時間改變。官司了結時，祖父兩個兒子均已成婚（銘洽二子，在其兄婚後不久，亦舉行純中式的婚禮）(141)，祖父凡事稱心如意，對前途充滿信心。

繼承人

一九二八年四月三十日下午祖父如常赴裕記會所用膳時，途中遇刺身亡，年僅四十七歲。遺下一妻、三妾、七子、六女，幼女舜娥為四姨太遺腹女。南華體育會特下半旗對會長—祖父致哀，利舞台及利園均歇業四十九天以示對其創辦人哀悼。祖父遺體靈堂設於利園，供生前好友憑弔，照中國習俗死者遺體須由妻、妾及子孫陪靈四十九日，並由不同廟宇之和尚誦經超渡亡魂。

祖父生前在南北行中的華商圈素享盛譽，又是當時香港最大地主之一，風度翩翩，亦頗得當地外籍人士人緣。

在事後，家人懸賞一萬元捉拿兇手。據報導警方在五月一日於澳門逮捕一名疑兇，但謠傳無法証實，警方雖然毫無線索，但仍鍥而不捨查緝兇徒。曾有目擊者指出案發時見一男子及少年急忙離開現場，但又無法詳細描述該男子外貌。家人在警方辦案期間均極為焦慮，多人曾赴警局查詢案情進展。雖然曾多次謠傳警方已緝拿真兇歸案，但此案至今依然懸而未破[142]。

祖父訃文於五月廿四日刊登于南華早報，宣佈次日出殯。送葬行列於早上十時由堅尼地道七十四號喪宅出發，沿皇后大道東、皇后大道中，經銀行及消防局，經過皇后大道中二〇二號的家族公司禮昌隆號，至設於堅尼地城的停靈帳棚，供友人於下午一時至二時之間做最後弔祭。祖父葬於香港仔的華人永遠墳場。

慘劇發生之時，父親、母親、榮根叔父（孝和）、舜英、舜華姑姐均在英國讀書，噩耗傳到之後立刻全體趕回香港奔喪，由於當時往返港、英之間僅有航船，行程費時甚久，故無法趕上喪禮。

祖母是祖父遺囑指定的唯一執行人，但祖母對生意毫無概念，便完全依賴父親處理一切。由

於事出突然，當時年僅二十三歲的父親頓時成為一家之主。當務之急是需要籌足大筆現金，以支付祖父的遺產稅。故須拋售幾乎所有股票，以換取現金。其中以九龍電力公司股票最多，由Sir Elly Kadoorie購得，他購進股票不久即加入公司董事會(143)。這是父親生平第一次須低頭向祖父生前好友告貸，體驗到人情冷暖，祖父一位朋友甚至吩咐家中下人，不可讓父親進門，父親永遠無法寬恕此人，連對他的家人都成反感(144)。在當時對父親伸出援手的朋友，此後便成為父親的世交摯友。

當時香港滙豐銀行會計主席Mr. Arthur Morse，挺身而出幫助父親渡過難關。當父親去銀行求見時，他立刻邀約「世侄」（他如此稱呼父親）共進午餐，父親對他的大力相助一直永誌在心(145)。

日後許多人向祖母出價，有意購買祖父的產業，惟在父親的鼓勵之下，祖母堅不變賣，然而父親須有能力償付房屋物業貸款，才能繼續保有產業。幸得Morse先生幫父親向香港滙豐銀行安排貸款，並在家人能力範圍之內分期按月償還。祖母與父親原本十分擔憂怡和洋行會再取回銅鑼灣產業(146)，但顯然此項財務安排十分成功。直到今日，香港滙豐銀行始終是利氏家族香港事業的往來銀行。

一九四一年日軍佔領香港後，市區情況一片混亂，所幸有香港滙豐銀行代理貸款財務。家人在赴內陸避難時，可將土地物業契據証件在銀行保險箱妥為保存，免在暴亂中遺失(147)。

祖父死後每月分期償還貸款對家人來說非常吃力。當時我們有大批土地，但週轉現金短絀，要由莫光先生為我們定時向租客收租。身為家中女家長的祖母不但配給每房家人家用，甚至祖父的兄弟，在生活最艱難的時候亦一樣家用無缺(148)，從未發生任何爭議。

今天我們所有子孫對祖母當日堅不變賣任何產業的決心由衷感激。多年來房地產業一直為家族的主要企業，並促使希慎興業有限公司的成立，由父親出任主席，更在一九八一年公開上市。

祖父對社區公益及教育機構熱心支持的精神，繼續由其子孫發揚光大，而希慎興業有限公司，今日則由其長孫利漢釗兄主政。